Energy Explorations: Sound, Light, and Heat

Developed and Published
by
AIMS Education Foundation

This book contains materials developed by the AIMS Education Foundation. **AIMS** (**A**ctivities **I**ntegrating **M**athematics and **S**cience) began in 1981 with a grant from the National Science Foundation. The non-profit AIMS Education Foundation publishes hands-on instructional materials that build conceptual understanding. The foundation also sponsors a national program of professional development through which educators may gain expertise in teaching math and science.

Copyright © 2010 by the AIMS Education Foundation

All rights reserved. No part of this book or associated digital media may be reproduced or transmitted in any form or by any means—including photocopying, taping, or information storage/retrieval systems—except as noted below.

- A person or school purchasing this AIMS publication is hereby granted permission to make up to 200 copies of any portion of it (or the files on the accompanying disc), provided these copies will be used for educational purposes and only at one school site. The files on the accompanying disc may not be altered by any means.

- Workshop or conference presenters may make one copy of any portion of a purchased activity for each participant, with a limit of five activities per workshop or conference session.

- All copies must bear the AIMS Education Foundation copyright information.

AIMS users may purchase unlimited duplication rights for making more than 200 copies, for use at more than one school site, or for use on the Internet. Contact us or visit the AIMS website for complete details.

AIMS Education Foundation
P.O. Box 8120, Fresno, CA 93747-8120 • 888.733.2467 • aimsedu.org

ISBN 978-1-60519-041-9

Printed in the United States of America

I HEAR AND
I FORGET,

I SEE AND
I REMEMBER,

I DO AND
I UNDERSTAND.

-Chinese Proverb

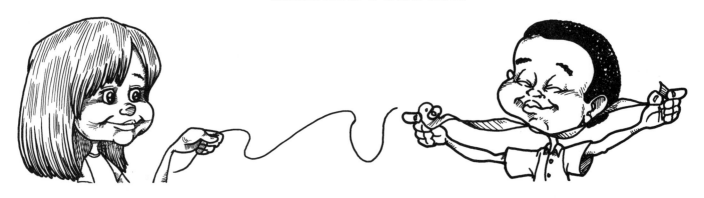

Energy Explorations: Sound, Light, and Heat

Table of Contents

Assembling Rubber Band Books

Rubber band books offer valuable content information in a kid-friendly way. Each student can be given his or her own book to keep and refer to at a later date. These books also provide a great home link, as students can take them home and share the information they are learning with their parents. To assemble a book, follow these simple instructions:

Fold back

Then over

Nest together

Hold together with a large rubber band

A #19 rubber band fits perfectly. If these are not available, snip the top and bottom of the center fold line of the book so that the other rubber bands can fit.

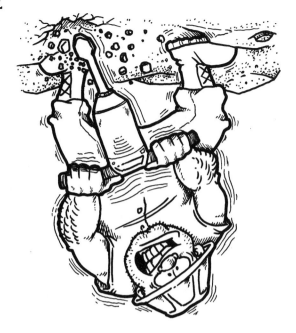

Sound waves carry energy through air (or other materials) as the molecules in them are pushed and pulled by a vibrating source. All sounds begin when something vibrates.

Energy is the ability to do work. It can cause motion and create change. Objects have energy and can gain energy or lose energy to other objects. A moving car has energy. A pot of water heating on a burner is gaining energy from the burner. A bowling ball loses energy as it hits the pins.

Everything has energy. It is everywhere. It has many forms and it can change from one form to another.

FORMS OF ENERGY

What is energy? It is not matter; it does not have mass. You cannot hold it in your hand.

© 2010 AIMS Education Foundation

Plants use light energy from the sun to make food. They convert light energy into chemical energy. Humans and animals depend on the chemical energy stored in plants to operate their bodies and muscles.

A ceiling fan changes electrical energy into mechanical energy. Electricity powers the motor, and the motor turns the blades. The blades move the air to cool you down.

Heat energy is the random motion of molecules. Molecules in matter are always in motion, but the hotter something is, the faster the molecules move. Temperature is a measure of that motion.

Energy appears in many forms including **light**, **heat**, **sound**, **electrical**, **chemical**, and **mechanical**. Though each form is different, they are all the same in the fact that one form of energy can change into another.

© 2010 AIMS Education Foundation

Isn't It Interesting...
Traveling Tips

Light travels at the speed of 300,000 kilometers per second (186,000 miles per second). This means that light takes 1/2 millionth of a second to go from home plate to an outfield fence located a distance of 125 meters (410 feet) and only 0.02 of a second to go from Los Angeles to New York.

Sound travels at various speeds through different conditions. However, an approximate speed of one mile per five seconds can be used to determine the distance a lightning strike is from your location. Start counting seconds as soon as you see the lightning flash and quit counting when you hear the thunder. For every second you count, the lightning is about 1/5 of a mile away.

Because of the rotation of our planet, people standing on the Equator are moving faster than 1730 kilometers per hour (1000 miles per hour)! The rest of us who are not standing on the Equator are not moving quite as rapidly. We don't notice this speed because the motion is constant and has been going on all our lives.

Air is a good insulator because the particles in it are relatively far apart, making it difficult for them to bump into each other to conduct the heat. An even better insulator is a vacuum—nothing at all! If no particles exist, heat cannot travel by conduction. (Think about your thermos bottle.)

 © 2010 AIMS Education Foundation

Sounds can be high or low. The difference in the highness or lowness of sound is called pitch. Higher pitches are made when objects vibrate faster. Lower pitches are made when those objects vibrate slower.

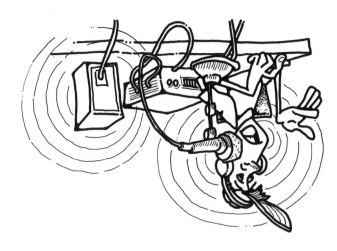

All sounds we hear have one thing in common. Sound is produced by vibrating objects. Vibrations are back-and-forth motions. These vibrating objects make sound waves. The waves go out in every direction. If you could see them, they would look like ripples that spread out when a pebble is dropped in a pond.

Put your fingers on your larynx. Hum a low pitch and feel the vibrations. Now hum a higher pitch and feel the vibrations. Humans can hear vibrations between 20 and 20,000 vibrations per second. Animals such as dogs, bats, dolphins, and whales can hear sounds with more vibrations.

HEAR THE VIBES

Sounds are all around us. We hear the wind moving through trees, the honking of car horns, friends laughing, music playing, and countless other sounds. Sounds never stop, not even when we sleep.

SOUND IS Vibration

Topic
Sound

Key Question
How is sound made?

Learning Goals
Students will:
- learn how sounds are produced,
- understand that vibrating objects vibrate whatever they touch, and
- observe that the length of the vibrating objects affects the pitch.

Guiding Document
NRC Standard
- *Sound is produced by vibrating objects. The pitch of the sound can be varied by changing the rate of vibration.*

Science
Physical science
    sound energy

Integrated Processes
Observing
Communicating
Collecting and recording data
Comparing and contrasting

Materials
Wooden rulers
Tuning forks
Craft sticks
Table tennis ball
Thread and tape
Rubber band
Pencil
Glass of water

Background Information
Energy must by used to produce sound. Whether it is the plucking of the strings of a guitar, the striking of a drum, or the blowing of a trumpet, energy is involved. The energy causes the object to vibrate, producing sound.

Whenever a sound is produced, something is quivering, throbbing, vibrating. Such movements are the basis of the sound we hear. Sound vibrations can be seen, like a violin string vibrating; they can be felt, like the vibration of a person's vocal cords; or they can be heard, like a ticking clock. All sounds can be traced to a vibration of some material.

In order for a sound to be heard, the vibrating material must move back and forth at least 16 times per second. The vibrating material may be a solid, liquid, or gas.

Management
1. Put materials in a location where students can get them as needed.
2. Tuning forks (item numbers 1973 and 1974) and table tennis balls (item number 1975) are available from AIMS.

Procedure
1. Begin by asking the students to close their eyes. Have them listen to all the sounds around them. Encourage them to describe the sounds they hear and judge whether the sound is high or low, etc.
2. Distribute the first student page. Have students record sounds they heard.
3. Take students outdoors. Have them listen and record the sounds they hear.
4. Go back to the classroom. Invite students to share the different sounds they heard.
5. Discuss that all sound is produced by vibrations that make sound waves.
6. Tell students that they will be investigating some sound-making devices. Distribute the second student page. Ask them to follow the directions on the page and make observations of the sounds that are produced.

Connecting Learning

1. What do all sounds have in common? [vibrations, vibrating sources, traveling energy]
2. What do vibrating objects look like? [objects moving back and forth rapidly, blurry]
3. What do vibrations feel like? [tingly, wiggly, etc.]
4. Describe the change in the sound when you move the pencil and change the length of the rubber band. [The sound becomes higher or lower. The rubber band vibrations are different.]
5. What caused the table tennis ball to bounce away from the tuning fork? [rapid back and forth movements of the tines of the tuning fork]
6. What sounds can you hear now? What is causing them? How is it that you can hear them?
7. What are you wondering now?

Extensions

1. Have the students use the lists of sounds they made to create a poem. For example, the form of the poem could be

   Sounds

   I hear   children giggling,
               cars honking,
               leaves rustling,
               people singing,
               special sounds
                       with my ears.

2. Have the students remember sounds that they enjoy hearing. List their responses on a chart labeled "I Like to Hear..." Have them remember sounds that they do not like to hear and add those to the list under "I Do Not Like to Hear..." Later categorize the list into high, low; loud, soft sounds.
3. Brainstorm some important sounds and how people should react to them. [fire alarm, fire engine and ambulance siren, school bell, smoke detector alarm, teacher's or police officer's whistle]

SOUND IS Vibration

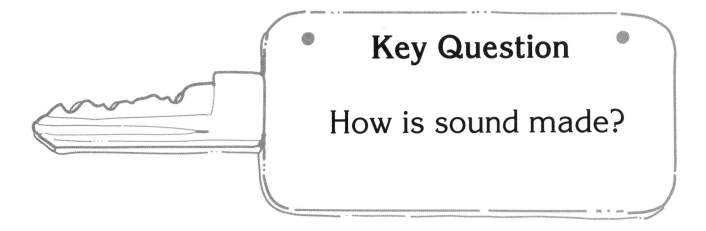

Key Question

How is sound made?

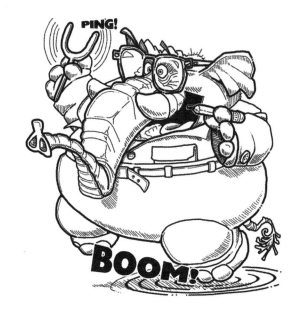

Learning Goals

Students will:

- learn how sounds are produced,

- understand that vibrating objects vibrate whatever they touch, and

- observe that the length of the vibrating objects affects the pitch.

                    © 2010 AIMS Education Foundation

((SOUND)) IS Vibration

Close your eyes and get very quiet. Listen to all the sounds you can hear in one minute. Open your eyes. Make a list of the sounds you heard.

1. _____
2. _____
3. _____
4. _____
5. _____
6. _____
7. _____
8. _____
9. _____

Go outside and sit quietly. Shut your eyes and listen for at least one minute. Open your eyes and make a list of the sounds you heard.

1. _____
2. _____
3. _____
4. _____
5. _____
6. _____
7. _____
8. _____
9. _____

How were the sounds different?

How were the sounds alike?

                    © 2010 AIMS Education Foundation

((SOUND)) IS Vibration

You will need:

craft stick
tuning fork
thread
table tennis ball
cup of water

rubber band
pencil
ruler
tape

Do this:

1. Hold one edge of a ruler tightly on your desk. Pluck the other end of the ruler and listen. Make the ruler shorter. Make the ruler longer.

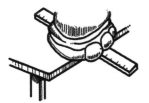  **shorter** **longer**

2. Clench a craft stick with your teeth. Pluck the end of the stick and listen. Change the length and try again.

3. Strike a tuning fork on your shoe (or hand). Watch and listen. Touch the vibrating tuning fork to the water's surface and watch.

4. Tape a length of thread to a table tennis ball. Touch the ball with the tines of a vibrating tuning fork and watch.

5. Touch your desk with the stem of a vibrating tuning fork and feel.

6. Make a mini-ukulele. Wrap a rubber band around a ruler. Push a pencil under the rubber band. Pluck the rubber band, watch, and listen. Move the pencil and try again.

7. How are all these activities alike?

 © 2010 AIMS Education Foundation

Connecting Learning

1. What do all sounds have in common?

2. What do vibrating objects look like?

3. What do vibrations feel like?

4. Describe the change in the sound when you move the pencil and change the length of the rubber band.

5. What caused the table tennis ball to bounce away from the tuning fork?

6. What sounds can you hear now? What is causing them? How is it that you can hear them?

7. What are you wondering now?

        © 2010 AIMS Education Foundation

Traveling Sounds

Topic
Sound

Key Question
Can sounds travel through solids, liquids, and gases?

Learning Goal
Students will compare and contrast sounds traveling through solids, liquids, and gases.

Guiding Document
NRC Standard
- *Sound is produced by vibrating objects. The pitch of the sound can be varied by changing the rate of vibration.*

Science
Physical science
   sound energy

Integrated Processes
Observing
Comparing and contrasting
Communicating
Predicting
Drawing conclusions

Materials
Wind-up clock
Large zipper-type plastic bag
Water
Table top
Empty metal coffee can
Paper bag
Glass jar
Shoebox
Old newspaper

Background Information
   Sound waves travel through every kind of material; the only place sound cannot travel is through a vacuum because there is nothing to vibrate.

Solids, liquids, and gases all conduct sound, but the speed of sound is different for each type of material. Most sounds that we hear are transmitted through air. Sound waves travel much faster through solids and liquids than through gases because the molecules of solids and liquids are closer together.

Sample Speeds

	meters per second	feet per second
Air	330	129
Water	1500	4794
Wood	4500	14,850
Metal	5000	16,500

Management
1. Fill the plastic bag with water and seal tightly.
2. Collect a coffee can, paper bag, glass jar, and shoebox with newspaper. They must be large enough to put the clock inside.
3. If a wind-up clock that ticks is not available, use a music box or hit two spoons together.

Procedure
1. Wind the clock. Hold it in the air. Have the students close their eyes and listen to the clock ticking.
2. Have students raise their hands if they can hear the ticking. Ask, "What is the sound traveling through?"
3. Explain that the air is a gas. Sound is produced by vibrations of the clock disturbing the air.
4. Have the students press their ear against a wood surface (table top, desk, wooden floor) and place the clock on the table top some distance from the students' ears. Ask, "Can you hear the sound?" "Is the sound louder or softer than when you heard it through air?" [louder because the wooden surface vibrates causing more air molecules to vibrate.]

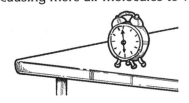

19   © 2010 AIMS Education Foundation

5. Have a student hold the water-filled bag to his/her ear. Hold the clock against the other side of the bag. Ask the student, "Can you hear the clock ticking?" [Yes.]
6. Put the clock inside each of the following in turn: a metal can, a paper bag, a glass jar, and a shoebox stuffed with newspaper. Have the students tell when the sound was the loudest and when it was softest. Which container(s) allowed you to hear the clock the loudest?
7. Try having the students press their ears against a metal surface (metal cabinet, metal door, or metal ruler) and put the clock on the metal surface. "Does sound travel through metal?" [Yes!]

Connecting Learning

1. Of the items supplied, which ones conducted sound?
2. Which did you think were the best conductors?
3. What surprised you in this lesson?
4. Scientists have recorded the sounds of whales and dolphins from miles away. How does this lesson apply to the ability to do this?
5. Name some instances when you would not want to be surrounded by good conductors of sound.
6. What other materials do you think would help to soften sounds? How could you investigate these materials?
7. Native Americans used to put their ears to the ground to listen for approaching buffalo herds. How does this lesson apply to this technique?
8. What are you wondering now?

Extensions

1. Take the students outside and have them press one of their ears to the ground. Bounce a ball or stomp your feet at least 10 feet away. Ask them if they heard you.
2. Have the students think of some different materials that sound can travel through and test each one. A student page is provided for this extension.
3. Bend the ends of a metal clothes hanger so that you can hold them to the outside of your ears with your fingers. Have someone tap the hanger *lightly* with a pencil. Listen while someone else tries it. Why does it sound so loud when you do it and so soft when someone else tries it? [The vibrations go through your skull bones when you do it, and through the air when someone else does it.]

20 © 2010 AIMS Education Foundation

TRAVELING SOUNDS

Key Question

How do sounds traveling through solids, liquids, and gas compare?

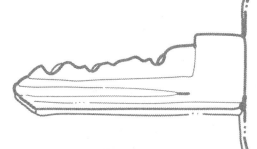

Learning Goal

Students will:

compare and contrast sounds traveling through solids, liquids, and gases.

Traveling Sounds

You will need:

glass jar · paper bag · metal can · shoebox · wind-up clock · table · newspaper · large plastic bag of water

Do this:

1. Listen to the clock. The sound is traveling through air.
2. Put the clock on a table. Press your ear to the table. The sound is traveling through wood.
3. Fill a plastic bag with water. Listen to the clock through it. The sound is traveling through water and plastic.

When was sound the loudest? Why?

air · wood · water

4. Put the clock inside a metal can, paper bag, glass jar, and a shoebox with newspaper.

When was the sound softest? Why?

metal can · paper bag · glass jar · shoebox with newspaper

TRAVELING SOUNDS

EXTENSION

What can sound travel through?

Make a list of your predictions. Test each one. Then make a new list of things you found that sound could travel through.

I think that sound can travel through these things.

I plan to test my predictions by

I found out that sound travels through these things.

 © 2010 AIMS Education Foundation

Traveling Sounds

Connecting Learning

1. Of the items supplied, which ones conducted sound?

2. Which did you think were the best conductors?

3. What surprised you in this lesson?

4. Scientists have recorded the sounds of whales and dolphins from miles away. How does this lesson apply to the ability to do this?

5. Name some instances when you would not want to be surrounded by good conductors of sound.

 © 2010 AIMS Education Foundation

Traveling Sounds

Connecting Learning

6. What other materials do you think would help to soften sounds? How could you investigate these materials?

7. Native Americans used to put their ears to the ground to listen for approaching buffalo herds. How does this lesson apply to this technique?

8. What are you wondering now?

SLINKY® SOUND

Topic
Sound

Key Question
How can we use string to conduct sound?

Learning Goals
Students will:
- compare and contrast different sounds that are made, and
- investigate that sound travels through solids.

Guiding Document
NRC Standard
- *Sound is produced by vibrating objects. The pitch of the sound can be varied by changing the rate of vibration.*

Science
Physical science
   sound energy

Integrated Processes
Observing
Comparing and contrasting
Communicating
Drawing conclusions

Materials
For each group:
   small plastic or paper cups
   paper clips
   metal Slinky® (see *Management 3*)
   metal fork or serving spoon
   metal coat hanger
   two lengths of string, 1 meter each

Background Information
This activity is an effective way to show that sound energy can be conducted through a stretched string better than through open air. The vibrating string can carry sound, and when the vibration is stopped, the sound is stopped.

If we want to hear a conversation that is taking place in the room next to us, we often place our ear on the door. We are hearing sound that probably traveled through the air and then through the door, making use of solids as sound conductors.

Management
1. This activity works well with students grouped in fours.
2. If you don't have enough forks and spoons, groups can share. High-quality silver serving forks and spoons have wonderful tonal qualities. If possible, have a variety of metals and sizes available for contrast.
3. Metal Slinkys® are available from AIMS (item number 1981).

Procedure
1. Ask the *Key Question* and state the *Learning Goals.*
2. Tell students that they are going to use string to listen to the sounds made with a metal Slinky®.
3. Show students how to tie the strings to the small plastic cups and then to the Slinky®. Inform them that two members of their group will need to hold the ends of the Slinky®, one member will hold the cups to his or her ears, and the fourth student will pluck or tap the Slinky®. Have students switch roles so all have a chance to listen to the Slinky®.
4. Discuss what the students observed. Allow them to share their surprise at how well the strings conducted the sounds.
5. Distribute a metal hanger to each group. Ask students to work in pairs. Have students tie the strings to the hangers and attach cups to the free ends. Invite them to tap the hangers, or to swing them so they hit the edge of the desk or back of the chair.
6. Discuss what the students observed.
7. Follow the same procedure for the forks and spoons.

Connecting Learning
1. What transmitted the sound from the Slinky® to your ears? [the string]
2. Could you see the string vibrating?
3. What other things would you like to try?
4. Name the things that vibrated when you used the Slinky®. ...the coat hanger. ...the silverware.
5. What are you wondering now?

© 2010 AIMS Education Foundation

SLINKY® SOUND

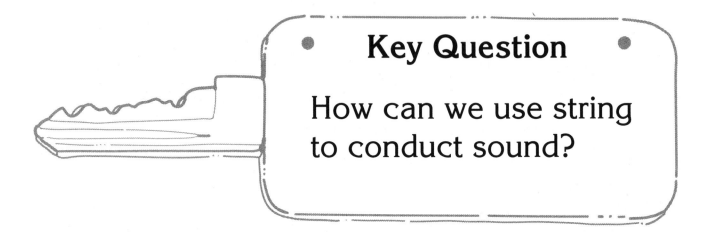

Key Question

How can we use string to conduct sound?

Learning Goals

Students will:

- compare and contrast different sounds that are made, and

- investigate that sound travels through solids.

SLINKY® SOUND

You will need:

2 paper or plastic cups

1 meter of string

2 paper clips

1 metal Slinky®

Do this:

1. Make a hole in the bottom of each cup.
2. Cut the string in half. Insert one end of each string into each of the cups.
3. Tie each string to a paper clip inside each cup.
4. Stretch the Slinky® to about 3 meters in length.
5. Tie the end of one of the strings to the middle to the Slinky®. Tie the other string near the first one.
6. Put both cups to your ears. Step away from the Slinky® until the strings are pulled tight. Have a partner pluck one end of the Slinky®.
7. Listen! What does it sound like to you?

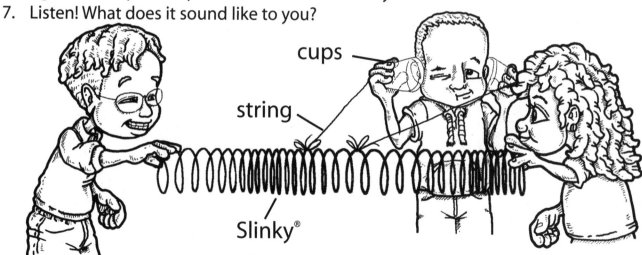

cups

string

Slinky®

8. Try a coat hanger, fork, or spoon.

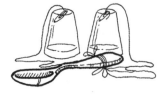

© 2010 AIMS Education Foundation

SLINKY® SOUND

Connecting Learning

1. What transmitted the sound from the Slinky® to your ears?

2. Could you see the string vibrating?

3. What other things would you like to try?

4. Name the things that vibrated when you used the Slinky®. …coat hanger. …silverware.

5. What are you wondering now?

 © 2010 AIMS Education Foundation

Topic
Sound

Key Question
How can we use cups to make sounds louder?

Learning Goal
Students will compare the sounds of plucked rubber bands and scissors stroking a metal Slinky® without cups and with cups as sounding boards.

Guiding Document
NRC Standard
• *Sound is produced by vibrating objects. The pitch of the sound can be varied by changing the rate of vibration.*

Science
Physical science
   sound energy
      amplification

Integrated Processes
Observing
Comparing and contrasting
Inferring

Materials
Metal Slinkys®
Plastic cups, 10 oz
Plastic cups, 9 oz
Rubber bands, various sizes
Scissors, metal

Background Information
Sound is produced when something vibrates. Vibrations are back and forth movements. Sound waves can travel through all kinds of material; the only place sound cannot travel is through a vacuum because there is nothing to vibrate.

Solids, liquids, and gases all conduct sound, but the speed of sound is different for each type of material. Most sounds that we hear are transmitted through air. Sound waves travel much faster through solids and liquids than through gases because the molecules of solids and liquids are closer together.

In this activity, students will observe the difference in the loudness of sound as it travels through air and as it travels through a solid. In the first experience, students will stretch a rubber band between their forefinger and thumb. Another student will pluck the rubber band as they listen to the sound that is produced. Students will then wrap the rubber bands around a plastic cup and pluck them to compare the sounds. In the second experience, students will run their scissors across the Slinky®. The sound heard as the waves pass through the air is quite faint. When a plastic cup is placed in both ends of the Slinky® and it is again stroked with the scissors, the resulting sound is quite loud (and futuristic!). The vibrations are transferred to the cup, a much larger surface, which in turn vibrates more air. The cup serves the same purpose that a sounding board in a piano serves. The strings in a piano force the sounding board to vibrate. Its larger surface produces a louder sound.

Management
1. Each group of students will need four rubber bands (at least two different sizes) and a cup of each size. Test the rubber bands beforehand to make certain that they will go around the cup from top to bottom.

2. Metal Slinkys® are available from AIMS (item number 1981).
3. Have students work together in groups of four. You will need a Slinky® for each group. If this is not possible, have one group demonstrate for the class, then put the Slinky® and the cups at a center for other groups to explore.
4. If possible, show the students the strings and sounding board in a piano. Also show them the mechanism and sounding board in a music box.

 © 2010 AIMS Education Foundation

Procedure

1. Ask the *Key Question* and state the *Learning Goal.*
2. Tell the students to be as quiet as possible. Wrap a rubber band around your forefinger and your thumb. Stretch it tight and pluck it. Ask how many heard it. Discuss the loudness of the sound.
3. Ask students how the sound could be amplified—made louder. Take their suggestions and, if possible, try them.
4. Distribute the rubber bands to each group of four students. Have them stretch the rubber bands between their fingers and thumbs and pluck.
5. Give each group two plastic cups, one of each size. Instruct the students to place the rubber bands so they go across the bottom and the top of the cup. Have students pluck them. Discuss the sounds that they hear. Are they louder? Why? [The rubber band vibrates the solid surface of the cup. The cup then vibrates the air around it. A larger surface vibrating = more air vibrating.] Did the rubber bands all have the same pitch? [No.] What made the difference? [its width and the amount it was stretched]
6. Ask students what musical instrument(s) their cups are like. [guitar, banjo, violin, cello]
7. Distribute the rubber band book. Allow students time to read it.
8. Distribute a Slinky® to each group. Have two students hold the Slinky® so that they are standing 10 to 15 feet apart.
9. Invite the other two students in the group to stroke Slinky® with their scissors. Talk about how loud the sound is.
10. Distribute two 10-oz plastic cups to each group. Have the two students holding the Slinky® place a cup in each end. Again invite the other two students to stroke the Slinky® with their scissors.
11. Discuss the effect the cups had on the loudness of the sound.

Connecting Learning

1. When we plucked the rubber bands held between our fingers, what vibrated so we could hear the sound? [the air]
2. When we put the rubber bands around the cups and plucked them, what vibrated? [the cups and the air]
3. Why was the sound amplified (louder)? [There was a larger surface (the cup) that vibrated more air.]
4. Describe what we did with the Slinky®.
5. How was it like the cups and rubber bands?
6. How was it different?
7. What are you wondering now?

Extension

Try plastic milk jugs in place of the cups at the ends of the Slinky®.

AMPLICUPS

Key Question

How can we use cups to make sounds louder?

Learning Goal

Students will:

compare the sounds of plucked rubber bands and scissors stroking a metal Slinky® without cups and with cups as sounding boards.

AMPLICUPS

Hold your rubber band like this and pluck it.

What vibrated?

Now wrap your rubber bands around a cup like this:

Pluck the rubber band. What vibrated?

Why was the sound louder?

How did the cups in the Slinky® change the loudness of its sound?

What vibrated?

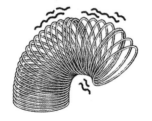

The word amplified means to make louder. Why do you think this lesson is called *Amplicups?*

We can describe sounds in many ways. One of those ways is to say whether it is loud or soft. How do you make a sound louder?

If you are watching television, you turn up the volume. If you are listening to music in the car, you turn up the volume on the radio. What happens when you turn up the volume?

Be careful around loud sounds. They can damage your ears.

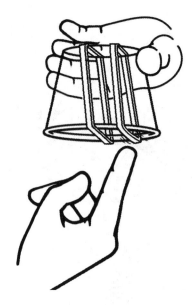

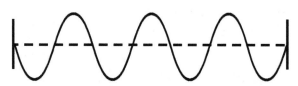

A louder sound wave looks like this. It has more energy.

A sound wave looks like this.

You just plucked some rubber bands on a cup. The cup made the sound louder. Why? Because the rubber bands make the cup vibrate. The cup has a larger surface that vibrated the air around it.

Which of these sound waves would represent the plucked rubber band on your fingers? Which would represent the plucked rubber band on the cup?

Have you ever felt the speakers vibrate? The higher the volume, the more the speakers vibrate. The more the speakers vibrate, the more they vibrate the air around them.

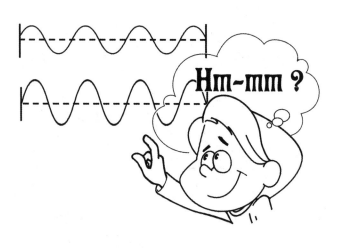

Hm-mm ?

© 2010 AIMS Education Foundation

Connecting Learning

1. When we plucked the rubber bands held between our fingers, what vibrated so we could hear the sound?

2. When we put the rubber bands around the cups and plucked them, what vibrated?

3. Why was the sound amplified (louder)?

4. Describe what we did with the Slinky®.

5. How was it like the cups and rubber bands?

6. How was it different?

7. What are you wondering now?

 © 2010 AIMS Education Foundation

CROWING CUPS

Topic
Sound

Key Question
What makes the cup crow?

Learning Goal
Students will make the chicken crow by pulling on its string to set up a vibration.

Guiding Document
NCR Standard
• *Sound is produced by vibrating objects. The pitch of the sound can be varied by changing the rate of vibration.*

Science
Physical science
    sound energy

Integrated Processes
Observing
Applying

Materials
For each student:
    red paper
    yellow paper
    white paper or googly eyes (see *Management 5*)
    plastic cup (see *Management 1*)
    string (see *Management 2*)
    paper clip
    sponges, optional (see *Management 7*)

For the class:
    scissors
    tape or glue
    pushpin

Background Information
All sounds are produced by vibrations. Your voice produces sounds because your vocal cords vibrate. Put your hand on some stereo speakers, turn up the volume, and feel the vibrations. A bell rings because its clapper hits its brass side causing it to vibrate. You hear the notes of a piano because the keys you strike vibrate strings.

In this activity, students will use their fingers to pull on a string causing it to vibrate. The cup serves to amplify the sound. By using short, quick pulls on the string, it sounds like a chicken.

Management
1. Cups of various sizes can be used. For small hands, 3-oz and 5-oz cups are easier to manipulate.
2. Cut the string into lengths about 15 to 18 inches long.
3. Prior to doing the activity, make a hole in the bottom of each cup with a pushpin. Turn the cup upside down, push the pin through the center of the bottom, and wiggle the pushpin around to enlarge the hole.
4. Thread the string from the bottom to the inside of the cup.
5. Eyes can be cut from white paper. Students can put black dots in them for the pupils. Googly eyes can be used instead of paper ones. Googly eyes can be purchased in sewing or craft departments.
6. The red and yellow paper is for students to cut out beaks and combs for their cups. No pattern is provided because the size of the cups used will determine the size that the beaks and combs need to be. Students can fold over a small portion of the shape they cut to tape or glue to the cup in the appropriate place.
7. It's easier for students to make the string vibrate if it's wet. You can tie a small cube of sponge to the bottom of the string. Dampen the sponge, pinch it around the string, and slide it down the string.

Procedure
1. Tell students they will make a chicken that crows.
2. Distribute the construction page and materials to each student.
3. Guide them through the construction process.
4. Demonstrate how to tug on the string to make the chicken crow.
5. Discuss the results.

Connecting Learning
1. What made your chicken crow?
2. Did all the cups sound the same? Explain.
3. What do you think would happen if we used a paper cup instead of a plastic cup? What would we have to do to find out?
4. What are you wondering now?

Key Question

What makes the cup crow?

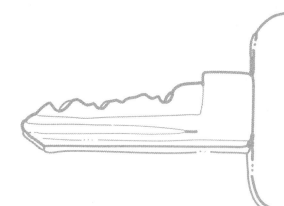

Learning Goal

make the chicken crow by pulling on its string to set up a vibration.

   © 2010 AIMS Education Foundation

"SQUAWK!" CROWING CUPS ASSEMBLY "SQUAWK!"

1. Tie a paper clip to one end of your string.

2. Push the string down through the hole in the bottom of the cup.

3. Cut a beak from yellow paper. Tape it on the cup.

4. Cut a comb from red paper. Tape it on the cup.

5. Make eyes for your chicken. Glue them on the cup.

6. Slide your fingernails down the string to make the chicken crow.

© 2010 AIMS Education Foundation

CROWING CUPS

"SQUAWK!" "SQUAWK!"

Connecting Learning

1. What made your chicken crow?

2. Did all the cups sound the same?

3. What do you think would happen if we used a paper cup instead of a plastic cup? What would we have to do to find out?

4. What are you wondering now?

 © 2010 AIMS Education Foundation

Musical Bottles

Topic
Sound

Key Question
How do different amounts of water affect the sound made by tapping the bottles?

Learning Goals
Students will:
• note the different pitches produced when they tap bottles filled with differing levels of water, and
• observe what happens when air is blown across the openings of the bottles.

Guiding Document
NRC Standard
• *Sound is produced by vibrating objects. The pitch of the sound can be varied by changing the rate of vibration.*

Science
Physical science
    sound energy

Integrated Processes
Observing
Predicting
Communicating
Collecting and recording data
Comparing and contrasting
Drawing conclusions

Materials
Five identical glass bottles (see *Management 1*)
Food coloring—red, blue, yellow, green, orange
Prediction graph
Colored markers
Pattern strip page
Glue
6" x 18" construction paper strips
Toy xylophone, optional

Background Information
Sounds are produced when things vibrate. The frequency of the vibration determines the pitch—how low or high something sounds. A number of variables affect pitch including such things as the material vibrating and its length. In a xylophone, the longer bars vibrate more slowly and have a lower pitch; the shorter bars vibrate more rapidly and have a higher pitch. When a bottle containing water is tapped, the glass and the column of water vibrate and make a sound. By adding different levels of water to the bottles, different pitches are created. The bottle with the least amount of water will vibrate the fastest and have the highest pitch. The one with the most water will vibrate the slowest and have the lowest pitch.

Note that the above situation is the opposite of blowing across the mouth of a bottle to produce a sound—much like one plays a flute. In blowing, it is the column of air that is set in vibration. Less full bottles have longer columns of air to vibrate and have lower pitches; fuller bottles have shorter columns of air and produce higher pitches.

Management
1. The five glass bottles needed for this activity are critical and must be tested before doing the activity with students. These bottles should be identical and produce a nice sound when filled with water and tapped. Bottles with smooth, straight sides that taper to a narrow opening work best. Narrower diameter bottles tend to produce nicer sounds than ones with wider diameters.
2. Using the graphing page as a guide, fill each bottle with a different level of water. Test the bottles by tapping them with a pencil. Add or pour out water until the pitches are evenly spaced.
3. Using the same page as a reference, add food coloring to the bottles as indicated. If orange food coloring is not available, mix red and yellow to make the orange.

4. Prepare the prediction graph by taping the two pages together. Color the bottles to match the color names.

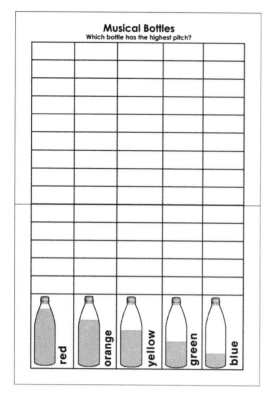

Musical Bottles
Which bottle has the highest pitch?

red   orange   yellow   green   blue

5. Prepare a song pattern strip for one or more simple songs. For example, *Mary Had a Little Lamb* would be yellow, orange, red, orange, yellow, yellow, yellow, orange, orange, orange, yellow, blue, blue. *Row, Row, Row Your Boat* would be red, red, red, orange, yellow, yellow, orange, yellow, green, blue.

Song Strip

Row, Row, Row Your Boat

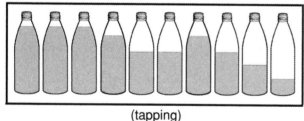

(tapping)

(blowing)

Procedure

1. If available, show the students a toy xylophone and ask them to describe it. Its bars may be painted with different colors that students notice, but keep probing until someone notes the different lengths of the metal plates. Play the xylophone to demonstrate the various pitches produced. Let students note that the shorter bars produce higher pitches while the longer bars produce lower pitches.

2. Place the five bottles on a table where students can easily see them. When viewed by students, the bottle with the most water should be on the left and the others should be placed in descending order to the right.

3. Show students the prediction graph and ask them to predict which bottle will make the highest sound or pitch when tapped. Have the students use the markers to record their predictions on the graph.

4. Let students come up and tap the bottles using a pencil. Repeat this process until students realize that the bottle with the most water has the lowest pitch and bottle with the least water has the highest pitch. Compare students' predictions to what actually happened.

5. Discuss ways the bottle xylophone and the xylophone are alike and different. Students should note that both instruments produce higher pitches with shorter vibrating materials (columns of water or steel bars) and lower pitches with longer vibrating materials.

6. Show students the song pattern strip and demonstrate how to use it to play a simple tune. Invite students to play the tune.

7. Distribute the pattern strip pages, scissors, construction paper strips, and glue. Let students color and cut out the bottle pictures to match the colors and water levels of the bottles in the bottle xylophone.

8. Have students create simple repetitive patterns using the pictures and glue them to the construction paper strips. When their strips are completed, let them come up and play their tunes on the bottle xylophone. For example, students might make a simple AAB pattern using blue and green and play the bottles accordingly (blue, blue, green; blue, blue, green; ...).

9. Ask students if there might be another way to make a sound with the bottles besides tapping them. Show them how blowing across the bottle openings produces a flutelike sound. Help students realize that in this case it is the air in the bottles that is vibrating. Thus, the bottle with the least air (most water) will have the highest pitch and the one with the most air (least water) will have the

          © 2010 AIMS Education Foundation

lowest pitch. This is exactly the opposite of what happens when the bottles are tapped.

10. End the activity with a class discussion.

Connecting Learning

1. Which bottle had the highest sound when it was tapped? [the one with the least water]

2. Which bottle had the lowest sound when it was tapped? [the one with the most water]

3. How is the bottle xylophone like the toy xylophone? [Various: Both are played by tapping. The longer bars have lower pitch and the longer columns of water do, too. Etc.]

4. How are they different? [Various: One uses water. The other is made of steel. One lays flat while the other stands up. Etc.]

5. How did your predictions compare to what actually happened?

6. What happened when air was blown across the opening of the bottle with the most water? [It produced the highest pitch.]

7. What happened when air was blown across the opening of the bottle with the least water? [It produced the lowest pitch.]

8. How was this different from what happened when the bottles were tapped? [The pitches are the opposite.]

9. Why does this happen? [When the bottles are tapped, the column of water vibrates. When the air is blown across their openings, the column of air vibrates.]

10. What are you wondering now?

Musical Bottles

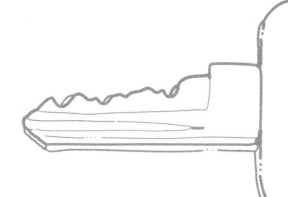

Key Question

How do different amounts of water affect the sound made by tapping the bottles?

Learning Goals

Students will:

- note the different pitches produced when they tap bottles filled with differing levels of water, and
- observe what happens when air is blown across the openings of the bottles.

46 © 2010 AIMS Education Foundation

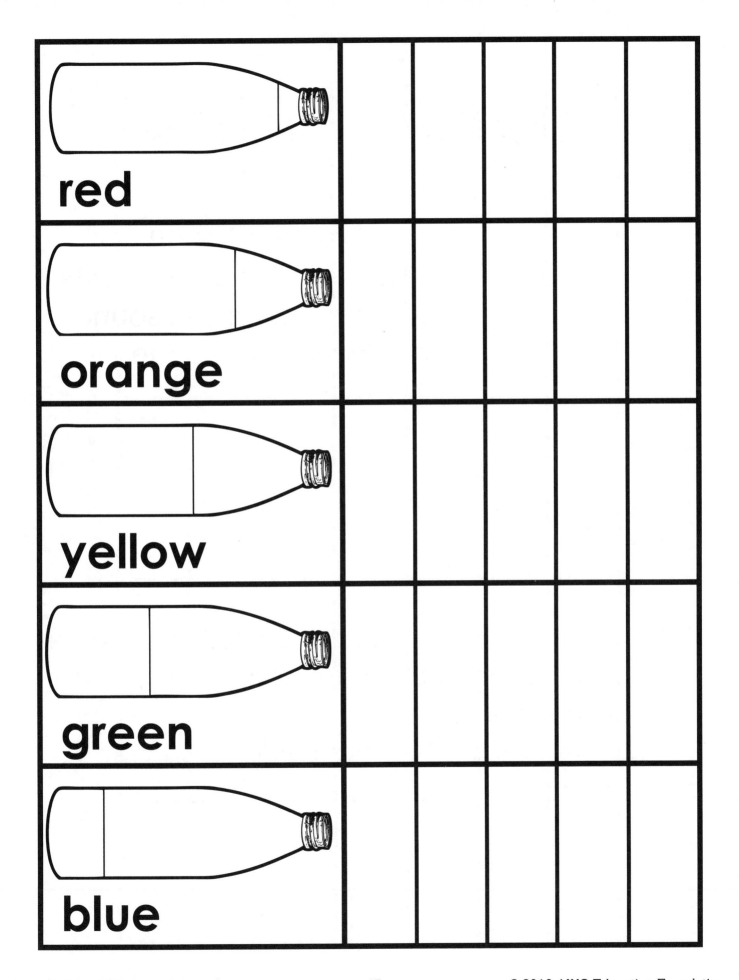

red

orange

yellow

green

blue

© 2010 AIMS Education Foundation

Musical Bottles

Which bottle has the highest pitch?

Musical Bottles

Pattern Strips

Musical Bottles

Connecting Learning

1. Which bottle had the highest sound when it was tapped?

2. Which bottle had the lowest sound when it was tapped?

3. How is the bottle xylophone like the toy xylophone?

4. How are they different?

5. How did your predictions compare to what actually happened?

6. What happened when air was blown across the opening of the bottle with the most water?

Musical Bottles

Connecting Learning

7. What happened when air was blown across the opening of the bottle with the least water?

8. How was this different from what happened when the bottles were tapped?

9. Why does this happen?

10. What are you wondering now?

Topic
Sound

Key Question
How can you make music with the Tune Thumpers?

Learning Goals
Students will:
- experiment with different ways to cause plastic tubes to make noise,
- order the tubes from lowest to highest pitch,
- recognize the relationship of a tube's length to its pitch,
- play tunes by thumping plastic tubes, and
- recognize that vibration is causing the sound.

Guiding Document
NRC Standard
- *Sound is produced by vibrating objects. The pitch of the sound can be varied by changing the rate of vibration.*

Science
Physical science
    sound energy

Integrated Processes
Observing
Comparing and contrasting
Analyzing
Generalizing

Materials
Fluorescent bulb covers (see *Management 1)*
Meter stick
Paper cutter
Permanent marker
Note cards (see *Management 4)*

Background Information
Sound is the result of vibration. The Tune Thumpers used in this activity are plastic tubes cut to specific lengths so that, when hit, they correspond to the notes of the C major scale. When the Tune Thumpers are hit, both the plastic tube and the air inside the tube vibrate, causing the sound that reaches our ears. The pitch of the sound changes with the rate of vibration (frequency). The faster the vibrations, the higher the pitch. The frequency of the vibrations is controlled by the length of the tube. The longer the tube, the slower the vibrations and the lower the resulting sound.

Management
1. You will need to purchase fluorescent light bulb covers to make the Tune Thumpers. These clear plastic tubes, about four centimeters in diameter, can be purchased from any home improvement store in eight- and four-foot lengths. One set of Tune Thumpers requires two eight-foot tubes or three four-foot tubes.
2. It is recommended that you make enough sets of Tune Thumpers so that each student in your class can have one tube. One set contains eight tubes.
3. The length of the tubes is critical to producing the correct pitch. Use a permanent marker to mark the light bulb covers at the appropriate increments and cut straight across with a paper cutter. Once a tube is cut, use the marker to identify the note (C, D, E, etc.). If desired, you can include the sol-fa syllables (do, re, mi, etc.) in addition to the letter.
    Middle C (Do): 62.7 cm
    D (Re): 55.6 cm
    E (Mi): 49.8 cm
    F (Fa): 46.3 cm
    G (Sol): 41 cm
    A (La): 36.3 cm
    B (Ti): 32.1 cm
    High C (Do): 30 cm
4. Copy the pages of notes onto card stock and cut them in half. If desired, make multiple copies of the notes for *Part Two* so that you can put the notes for an entire song across the board.

Procedure
Part One
1. Show the students a set of Tune Thumpers. Ask them if they think you could play a song using the tubes. Encourage them to share ideas about how the tubes could be used to make noise.
2. Invite several students to come to the front to try different ideas for ways to cause the tubes to make sound.

 © 2010 AIMS Education Foundation

3. Divide the students into groups and give each group a set of Tune Thumpers. Allow a time of free exploration during which students can experiment with ways to get the best sound from the tubes.

4. Have groups share the ways that they were able to produce sound and discuss which methods produce the nicest notes. Be sure they include dropping the tubes on the floor, hitting the tubes against the heel of a hand, hitting the tubes against each other, hitting them on the floor, and hitting them on a desk or other solid object.

5. Challenge each group to put the tubes in order from lowest to highest note (pitch). Have each group play their tubes in the order they have selected and see if the class agrees that the notes go from lowest to highest.

6. Once the tubes are in the correct order, invite students to make observations about the tubes and the notes they produce. Lead students to the recognition that the longer the tube, the lower the sound.

7. Place the half-page copies of the notes on a table. Challenge the class to arrange them in order from lowest to highest and place them in the board tray so they are visible to the whole class.

8. Discuss why the tubes make sound. Invite students to feel the vibrations, see the vibrations, and hear what happens when the vibrations are muffled.

Part Two

1. Explain that you are going to work together as a class to play a song. Make sure that each student has a tube. Group the students so that those with the same note are standing together and arrange them in order from low to high pitch.

2. Agree on how you will hit the Tune Thumpers to cause them to vibrate (on the floor, on the hand, etc.). Have students practice playing in unison when you call out the letter of the tube that they are holding. Go up and down the scale several times, pointing at the letters in the board tray as you call them out, until students are comfortable playing together on command.

3. Write the notes to one of the songs provided on the board (or line up the half-page copies of the notes along the board tray). Tell students that you will practice until they can play the song. Conduct the song by calling out the notes in sequence for students to play.

4. Repeat with other songs as desired. Be sure to select songs that will allow all students to have a chance to play. (No one song uses all the notes.)

5. Close with a time of discussion where students articulate what they have learned about sound and vibrations.

Connecting Learning

1. What are some ways you can get the Tune Thumpers to make noise? [dropping the tubes on the floor, hitting them against the heel of a hand, hitting them against each other, hitting them on the floor, hitting them on a desk, etc.]

2. Which method do you like best? Why?

3. Which Tune Thumper plays the highest note? [high C] …the lowest note? [middle C]

4. Describe the difference between a tube that makes a low sound and a tube that makes a high sound. [The tube that makes a low sound is longer.]

5. When you put the tubes in order from highest to lowest sound, what do you notice? [The length of the tube changes as the note changes. Longer tubes produce lower notes and shorter tubes produce higher notes.]

6. How do you think the length of the tube relates to the sound of the note? [The longer the tube, the lower the note.]

7. Without playing them, which Tune Thumper will play a higher note—the A or the E? [A] How do you know? [The A tube is shorter than the E tube.]

8. Why do the Tune Thumpers make noise when you hit them? [vibration]

9. What other things make noise in the same way? [All sound results from vibration, even though sometimes it's not as easy to see or feel as it is with the Tune Thumpers.]

10. What are you wondering now?

Extensions

1. Allow students to find the notes for other songs and play those using the Tune Thumpers.

2. Practice playing chopsticks or some other simple song that requires chords by hitting the two tubes that make the chord together.

3. Get a copy of *The Sound of Music* and show the scene where Maria is teaching the children to sing using the sol-fa notes. Challenge students to play along with the singing.

Tune Thumpers

Key Question

How can you make music with the Tune Thumpers?

Learning Goals

Students will:

- experiment with different ways to cause plastic tubes to make noise,
- order the tubes from lowest to highest pitch,
- recognize the relationship of a tube's length to its pitch,
- play tunes by thumping plastic tubes, and
- recognize that vibration is causing the sound.

Following are the notes for some familiar songs. No one song uses all the notes, so be sure to select a variety so that each student will have the opportunity to play. In the notation, C_1 corresponds to middle C and C_2 corresponds to high C.

London Bridges

G A G F E F G D E F G G A G F E F G D G E C_1

Mary Had a Little Lamb

E D C_1 D E E E D D D E G G E D C_1 D E E E E D D E D C_1

Twinkle, Twinkle Little Star

C_1 C_1 G G A A G F F E E D D C_1 G G F F E E D G G F F E E D
C_1 C_1 G G A A G F F E E D D C_1

Clementine

F F F C_1 A A A F F A C_2 C_2 B A G G A B B A G A F F A G C_1
E G F

The Wheels on the Bus

C_1 F F F F A C_2 A F G G G E D C_1 C_1 F F F F A C_2 A F G C_1
C_1 F

When the Saints Go Marching In

F A B C_2 F A B C_2 F A B C_2 A F A G A G F F A C_2 C_2 C_2 B A B
C_2 A F G F

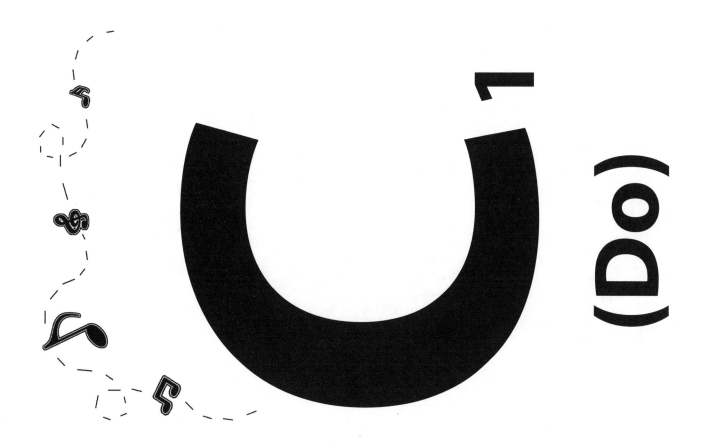

57

© 2010 AIMS Education Foundation

(Fa)

(Mi)

 © 2010 AIMS Education Foundation

2

(Do)

- - - - - - - - - - - - - - - -

(Ti)

© 2010 AIMS Education Foundation

Connecting Learning

1. What are some ways you can get the Tune Thumpers to make noise?

2. Which method do you like best? Why?

3. Which Tune Thumper plays the highest note? ...the lowest note?

4. Describe the difference between a tube that makes a low sound and a tube that makes a high sound.

5. When you put the tubes in order from highest to lowest sound, what do you notice?

Connecting Learning

6. How do you think the length of the tube relates to the sound of the note?

7. Without playing them, which Tune Thumper will play a higher note— the A or the E? How do you know?

8. Why do the Tune Thumpers make noise when you hit them?

9. What other things make noise in the same way?

10. What are you wondering now?

 © 2010 AIMS Education Foundation

♪ HUMDINGERS & Whistleblowers ♪

Topics
Sound, frequency, pitch

Key Question
What is the frequency of your vocal cords when you hum or whistle a note?

Learning Goals
Students will:
- realize that sounds are produced by vibrations;
- understand that the faster their vocal cords vibrate, the higher the pitch of the sound; and
- calculate the frequencies of notes on the piano and determine the frequencies of their vocal cords as they hum and whistle.

Guiding Documents
Project 2061 Benchmarks
- *Use numerical data in describing and comparing objects and events.*
- *Offer reasons for their findings and consider reasons suggested by others.*
- *Organize information in simple tables and graphs and identify relationships they reveal.*
- *Use numerical data in describing and comparing objects and events.*

NRC Standards
- *Sound is produced by vibrating objects. The pitch of the sound can be varied by changing the rate of vibration.*
- *Energy is a property of many substances and is associated with heat, light, electricity, mechanical motion, sound, nuclei, and the nature of a chemical. Energy is transferred in many ways.*
- *Mathematics is important in all aspects of scientific inquiry.*

NCTM Standards 2000*
- *Select appropriate methods and tools for computing with whole numbers from among mental computation, estimation, calculators, and paper and pencil according to the context and nature of the computation and use the selected method or tool*
- *Work flexibly with fractions, decimals, and percents to solve problems*

- *Recognize and apply mathematics in contexts outside of mathematics*
- *Develop and use strategies to estimate computations involving fractions and decimals in situations relevant to students' experience*

Math
Number and operations
    multiplication and division
Rational numbers
    decimals
Estimation
    rounding

Science
Physical science
    sound energy
      vibrations
      frequency

Integrated Processes
Observing
Comparing and contrasting
Classifying
Generalizing
Applying

Materials
Student page
Calculators
Access to piano or keyboard

Background Information
All sound is produced by vibrations. These vibrations can travel through air, through liquids, and through solids. Vibrations are back and forth movements. The number of back and forth movements (cycles) is called frequency. The greater the frequency, the higher the pitch. The slower the frequency, the lower the pitch. Sopranos and flutes have higher frequencies than do tenors and tubas.

Frequency can be expressed as the number of vibrations per second. The notes of musical instruments have distinct frequencies. On a piano, the note A above middle C has a frequency of 440 vibrations per second. The frequencies of all notes on the piano can be determined by calculating their values from this A. Actual frequencies for the notes are:

Note	Frequency (vibrations/sec.)
Middle C	261.63
C#	277.18
D	293.66
D#	311.13
E	329.63
F	349.23
F#	370.00
G	392.00
G#	415.31
A	440.00
A#	466.17
B	493.89

The notes in the next octave higher are determined by multiplying these frequencies by 2. For example: A is determined by multiplying the frequency for A above Middle C by 2 = 880.00. The notes in the next octave lower are these given values divided by 2. So A below the A above Middle C is 440 divided by 2 = 220.00. The frequencies for each subsequent higher octave are found by multiplying the frequencies in the adjacent octave by 2 or for subsequent lower octaves by dividing the frequencies in the adjacent octave by 2.

Student approximations will be determined by working upwards from A by multiplying its frequency by 1.059 and working downwards by dividing by 1.059. Use of a calculator is recommended. Students need to report results by rounding to the hundredths place.

Their approximations will be:

Note	Frequency (vibrations/sec.)
Middle C	261.54
C#	277.10
D	293.66
D#	311.13
E	329.63
F	349.23
F#	370.00
G	392.00
G#	415.31
A	440.00
A#	466.17
B	493.89

Management
1. Students will be asked to round decimal values to the nearest hundredths. Either review or explain how to do this if needed.

2. The use of calculators will make this activity move along faster.
3. Access to a piano is necessary so that students can find the notes they are singing or whistling. If a piano is not readily available, you can make a recording of the sounds or use a pitch pipe. Because a pitch pipe does not have the range that a keyboard or piano has, it carries less of an impact.
4. There are many virtual pianos available online if real pianos are not available. Two possible sites are http://www.virtualpiano.net and http://www.thevirtualpiano.com. You will need to help students identify Middle C and A on the pianos. Below is a portion of a keyboard with the keys marked.

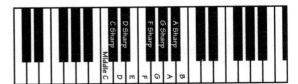

Procedure
Part One
1. Ask the *Key Question* and state the *Learning Goals*.
2. Invite students to lightly press their fingers against their larynxes. Tell them that the whole class is going to hum the song "Happy Birthday to You." Urge them to observe what is happening to their larynxes as they hum the song.
3. As students share their observations, write them on the board.
4. Tell them they are now going to whistle the song. Again, invite them to feel their larynxes as they do so. Record any new observations that students share.
5. After some class discussion, bring students to generalize that sounds are produced by vibrations. The notes they hummed are created by the vocal cords vibrating. Make certain that they understand that the different pitches are created by relaxing or tightening the vocal cords. They should have felt their larynxes bob up and down with the high notes and low notes they were humming.
6. Hand out the reading passage *Voice Box*. Discuss the passage and how it relates to their humming and whistling. If necessary, let students hum or whistle the Happy Birthday song again.

Part Two
1. Tell students that they are going to determine the number of vibrations their vocal cords produce to hum a note for one second. Ask them to make a guess, and write these guesses on the board.
2. Distribute the student pages. Tell students that they will first need to determine the frequency (vibrations per second) of the notes on a piano.

3. Point out the key for A above Middle C. Inform students that the piano wire for this note vibrates 440.00 times per second. They will use this number and their calculators to determine the frequency for the other notes in that octave.
4. Have them look at the instructions. Tell them that once they have figured the frequency for the octave that includes Middle C, they will multiply those results by two to determine the frequencies in the next octave higher and divide by two to determine the frequencies in the next octave lower.
5. When students are finished calculating, compare their results and discuss and rectify any differences.

Part Three
1. Invite the students to gather around the piano. Ask for a volunteer to hum or whistle a note. Match this note on the piano.
2. Have students refer to their calculations to determine the frequency of the vibrations of this note.
3. Ask for other volunteers, making sure to have a variety of pitches. Compare the note with the calculated frequencies.
4. Discuss how the greater frequencies make higher pitches, and the lower frequencies make lower pitches.

Connecting Learning
1. How did our guesses compare with the actual frequencies of the notes?
2. How does the frequency of a note determine the pitch? [Higher frequencies produce higher pitches.]
3. Which musical instrument has the greater frequency, a flute or a trombone? [flute] How do you know? [The flute plays higher notes. Higher pitches are created by greater frequencies.]
4. What happens to your larynx when you sing high notes? Why do you think this happens? [your larynx moves up, tightening the vocal cords, increasing their frequency]
5. As we age, our voices change—most noticeably with boys. How do they change? What do you think is happening? [As we grow and age, our vocal cords get longer. The vocal cords of boys get longer and heavier. Their voices will often crack because their muscles are having to get used to dealing with the larger vocal cords.]
6. What are you wondering now?

Extensions
1. Look at the piano strings. Compare the length and the thickness of those that produce high notes to those that produce low notes.
2. You can play a simplified version of Happy Birthday on the piano and have students compare the frequency for each note. Try this version: Middle C, Middle C, D, Middle C, F, E; Middle C, Middle C, D, Middle C, G, F; Middle C, Middle C, C, A, F, E, D; A Sharp, A Sharp, A, F, G, F.

* Reprinted with permission from *Principles and Standards for School Mathematics,* 2000 by the National Council of Teachers of Mathematics. All rights reserved.

 © 2010 AIMS Education Foundation

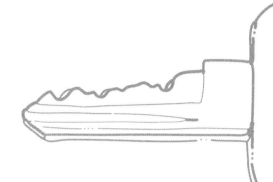

HUMDINGERS & Whistleblowers

Key Question

What is the frequency of your vocal cords when you hum or whistle a note?

Learning Goals

Students will:

- realize that sounds are produced by vibrations;
- understand that the faster their vocal cords vibrate, the higher the pitch of the sound; and
- calculate the frequencies of notes on the piano and determine the frequencies of their vocal cords as they hum and whistle.

Voice Box

When you breathe normally, the air goes through your nose into your windpipe. The windpipe is a hollow tube leading from the back of your throat toward the lungs. At the top of the windpipe is a hollow organ called the voice box or larynx. It is made of bone and cartilage held together by ligaments and muscles.

The vocal cords in your voice box are two, straight elastic-like strings. When you are not speaking or singing, these cords are relaxed against the sides of your voice box. When you start to talk or sing, tiny muscles bring the vocal cords closer together. Air coming up from the lungs makes the vocal cords vibrate back and forth very fast. When vocal cords vibrate, sounds are produced. Stretching the vocal cords tight and thin makes your voice high and squeaky. Relaxing them to a loose and wide shape makes low, deep sounds.

View looking down into larynx:

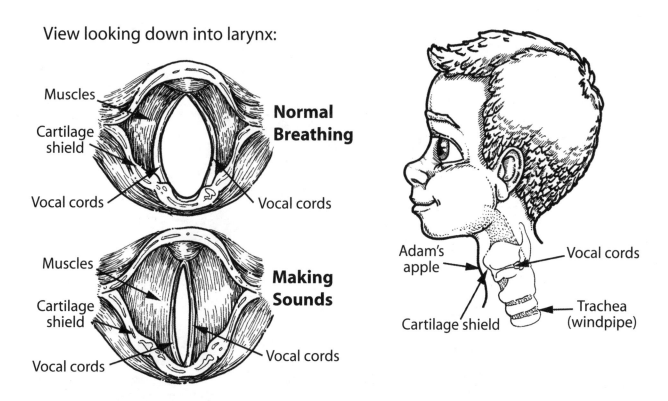

Muscles

Cartilage shield

Normal Breathing

Vocal cords

Vocal cords

Muscles

Cartilage shield

Making Sounds

Vocal cords

Vocal cords

Adam's apple

Vocal cords

Cartilage shield

Trachea (windpipe)

You can find your larynx just behind the bump in your throat called the Adam's apple. Feel around for the cartilage shield. The larynx is right behind it. Put your hand on your larynx and hum or talk. Feel the vibrations. Make the lowest sound that you can. Now make the highest sound. Feel the muscles move.

HUMDINGERS & Whistleblowers

What are the frequencies (vibrations per second) made by these notes on the piano?

The note A vibrates 440.00 times per second.
You can calculate all the other frequencies from there.

To determine the frequency for the note **above** A, which is A# (# is the symbol for the word sharp), you will need to **multiply** the frequency for A by 1.059. Get your calculator ready. So 440.00 x 1.059 gives you the frequency for A#. Round your answer to the nearest hundredth.

To find the frequency for the note B, you will multiply the answer for A# by 1.059. Round to the nearest hundredth.

To find the frequencies for the notes **below** A, you will **divide** by 1.059. So for G#, divide 440.00 by 1.059 and round to the nearest hundredth. For G, divide the answer for G# by 1.059 and round to the nearest hundredth. Just keep doing this until you are at Middle C.

Label your answers Hz. Hz stands for hertz, which means one vibration per second.

B	
A#	
A	440.00 Hz
G#	
G	
F#	
F	
E	
D#	
D	
C#	
Middle C	

♪ HUMDINGERS & Whistleblowers ♫

What are the frequencies (vibrations per second) made by these notes on the piano?

You've done the hard work. Now all you have to do is to multiply or divide by two to find the frequencies for all the other notes.

First, do you see how the names of the notes just keep repeating? Choose a note with a frequency you have already determined. Now locate the same note an octave higher. To find its frequency, multiply the frequency of the one you already know by two and label it Hz. That's it. Do the same for all notes above the ones you've done. For those below, just divide the frequencies you've figured by two and label them Hz. Record your answers on the next page.

Higher notes have _____ frequencies.
(higher or lower)

Lower notes have _____ frequencies.
(higher or lower)

Connecting Learning

1. How did our guesses compare with the actual frequencies of the notes?

2. How does the frequency of a note determine the pitch?

3. Which musical instrument has the greater frequency, a flute or a trombone? How do you know?

4. What happens to your larynx when you sing high notes? Why do you think this happens?

5. As we age, our voices change—most noticeably with boys. How do they change? What do you think is happening?

6. What are you wondering now?

Topics
Sound, pitch

Key Question
What makes the bug buzz?

Learning Goals
Students will:
- create a Buzzin' Bug,
- compare and contrast their bugs to determine what variables may cause the sounds,
- relate the frequency of vibrations to notes on the piano, and
- graph the frequency data.

Guiding Documents
Project 2061 Benchmarks
- *Use numerical data in describing and comparing objects and events.*
- *Offer reasons for their findings and consider reasons suggested by others.*
- *Organize information in simple tables and graphs and identify relationships they reveal.*
- *Use numerical data in describing and comparing objects and events.*

NRC Standards
- *Sound is produced by vibrating objects. The pitch of the sound can be varied by changing the rate of vibration.*
- *Energy is a property of many substances and is associated with heat, light, electricity, mechanical motion, sound, nuclei, and the nature of a chemical. Energy is transferred in many ways.*
- *Mathematics is important in all aspects of scientific inquiry.*

NCTM Standards 2000*
- *Recognize and apply mathematics in contexts outside of mathematics*
- *Collect data using observations, surveys, and experiments*
- *Represent data using tables and graphs such as line plots, bar graphs, and line graphs*

Math
Data collection
Ordering
Ranges
Graphing
  bar graph

Science
Physical science
  sound energy
    vibrations
    frequency

Integrated Processes
Observing
Comparing and contrasting
Classifying
Collecting and recording data
Interpreting data
Applying

Materials
For each student:
  2 wine corks (see *Management 2*)
  craft stick
  white glue
  10 cm x 10 cm piece of card stock
  1 m crochet thread or string
  rubber band (see *Management 3*)
  data sheet from *Humdingers and Whistleblowers*
  student pages

For the class:
  access to a piano or keyboard (see *Management 7*)

Background Information
All sound is produced by vibrations. These vibrations can travel through air, through liquids, and through solids. Vibrations are back and forth movements. The number of back and forth movements (cycles) is called frequency. The greater the frequency, the higher the pitch. The slower the frequency, the lower the pitch. Sopranos and flutes have higher frequencies than do tenors and tubas.

Frequency can be expressed as the number of vibrations per second. The notes of musical instruments have distinct frequencies. On a piano, the note A above middle C has a frequency of 440 vibrations per second. The frequencies of all notes on the piano can be determined by calculating their values from this A. Approximations due to rounding will be sufficient for this experience. Actual frequencies for the notes are:

Note	Frequency (Hz)
Middle C	261.63
C#	277.18
D	293.66
D#	311.13
E	329.63
F	349.23
F#	370.00
G	392.00
G#	415.31
A	440.00
A#	466.17
B	493.89

The "bugs" that students make in this activity will make a buzzing sound when twirled at the end of the string. By matching this note to a corresponding note on the piano, students will know the frequency of the vibrating rubber band on the bug. A real-world application can be made that will inform students as to the vibration of those pesky insects (mosquitoes and flies) that buzz around their ears; the pitch of their buzz is a result of the frequency of their wings. The buzz may not quite match a piano note, but students will be able to make a close approximation.

Management

1. Students must have completed the activity *Humdingers and Whistleblowers* before doing this one.
2. Several weeks ahead of doing this activity, ask parents to save corks. Corks can also be obtained from sources that supply home winemaking kits. They can be purchased at craft stores, but be aware that they will be expensive there.
3. The various widths and lengths of rubber bands will be at least one of the variables that causes different pitches of the bugs. You may want to start students off with the same size of rubber bands or let them discover the differences on their own by using various sizes. The rubber bands need to be at least 3 ½ inches long.
4. Students can twirl the Buzzin' Bugs in front of themselves or overhead. In any case, the twirling should be done in a location where students can spread out so that no one is hit.

5. If time permits, this activity is an excellent one for designing and re-designing.
6. The classification of the bugs comes from names of stereo speakers: *Woofers* are for bass sounds, *Tweeters* are for high-pitched sounds, and *Midranges* are for sounds in-between. Because this is a relative scale, students will need to come up with their own classification system. After determining the frequencies of their bugs, students can determine their own ranges for classifying. Differences in classifications will prompt interesting discussions.
7. Students will need access to a piano or keyboard in order to play the notes that match the buzz of their bugs. There are many virtual pianos available online if real pianos are not available. Two possible sites are http://www.virtualpiano.net and http://www.thevirtualpiano.com. You will need to help students identify Middle C and A on the pianos. Below is a portion of a keyboard with the keys marked.

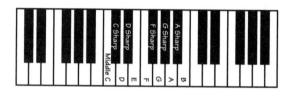

Procedure

Part One

1. Ask the *Key Question* and state the *Learning Goals*.
2. Distribute the materials and instructions for making the Buzzin' Bugs.
3. Tell students that they will not be able to twirl their bugs until the following day because the glue needs to completely dry so that the bug parts don't fall off.
4. Allow time for construction.

Part Two

1. Go to an area where students can twirl their bugs. Have them compare and contrast the sounds produced.
2. Have students find partners to help find the note on the piano that matches the buzz of the bug. Inform them that each pair of students will come to the piano. One student will twirl his or her bug while the other student matches the note on the piano. Other students can offer their advice as to whether the piano player should find a note that is higher or lower than the one tested. Once the note is determined, it should be compared to the data from the *Humdingers and Whistleblowers* sheet.
3. Have all students record the results for each bug on the *Buzzin' Bugs Data Sheet*.

4. After all data are collected and recorded, invite partners to work together to determine the frequency ranges for the categories of *Woofer, Mid-Range,* and *Tweeter.* They will need to order the frequencies from least to greatest and then decide on the ranges for the three categories. Have them color in the coding Key on the data page and then lightly shade the frequencies for each range. Ask the students to complete the bar graph using their frequency ranges.

5. Allow time for comparing and contrasting with an ensuing discussion of results.

Connecting Learning

1. What produced the sound of the Buzzin' Bug? [the vibration of the rubber band]

2. What is necessary for any sound to be made? [vibration]

3. Sounds do not travel in outer space. Why do you think this is? [Outer space is mostly a vacuum. There is no air for the sound waves to vibrate.]

4. What do you think made our Buzzin' Bugs sound differently? How do you know?

5. How did our graphs compare? What caused the differences?

6. Which category had the most bugs in it? How many more did it have than the category with the fewest bugs?

7. How many of you have had a mosquito buzz in your ear? What do you think produces the buzz?

8. How would you go about finding out how many times per second it beats its wings?

9. What other things hum or buzz? Could you determine what is vibrating and its frequency? How?

10. What are you wondering now?

* Reprinted with permission from *Principles and Standards for School Mathematics,* 2000 by the National Council of Teachers of Mathematics. All rights reserved.

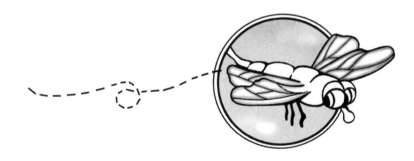

                    © 2010 AIMS Education Foundation

Buzzin' Bugs

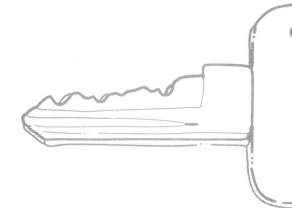

Key Question

What makes the bug buzz?

Learning Goals

Students will:

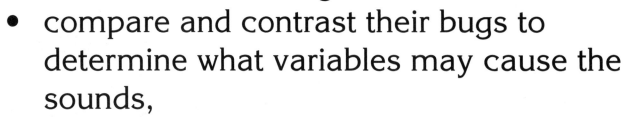

- create a Buzzin' Bug,
- compare and contrast their bugs to determine what variables may cause the sounds,
- relate the frequency of vibrations to notes on the piano, and
- graph the frequency data.

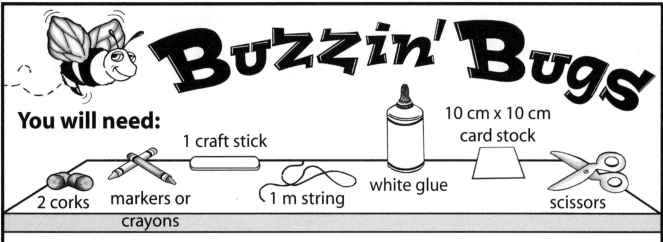

You will need:

2 corks • markers or crayons • 1 craft stick • 1 m string • white glue • 10 cm x 10 cm card stock • scissors

Do this:

1. Design an insect on the piece of card stock. Color it and cut it out.

2. Use the end of your scissors to make a slit in each cork.

3. Insert one end of your craft stick into the slit in one cork and the other end of the craft stick into the slit in the other cork.

4. Glue the areas where the craft sticks were inserted into the corks.

5. Put a line of glue along the craft stick and place your insect on it.

6. Put the Buzzin' Bug aside until the glue is dry.

Now do this:

1. Tie a string around one end of the craft stick. Make sure to double knot it.

2. Choose a rubber band to place around the two corks.

3. Go to an open area and stand at least two meters from anyone or anything. Twirl your Buzzin' Bug.

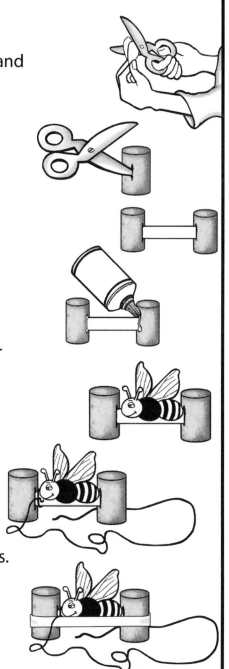

    © 2010 AIMS Education Foundation

Record the class data:

Name	Frequency	Name	Frequency

Key: ☐ Woofer ☐ Mid-Range ☐ Tweeter

Classify the Buzzin' Bugs into three ranges:
Woofers (those with low pitches)
Mid-Ranges (those with ranges between woofers and tweeters)
Tweeters (those with high pitches)

Record the range of pitches you decided for each category:

_____   _____   _____
        Woofers   Mid-Ranges   Tweeters

Complete the key on the data sheet. Color the data according to your classification. Now make a bar graph of your data.

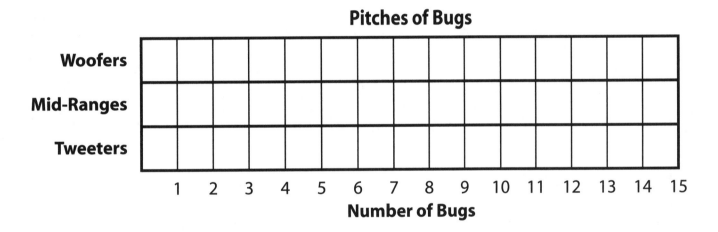

Pitches of Bugs

Woofers

Mid-Ranges

Tweeters

1 2 3 4 5 6 7 8 9 10 11 12 13 14 15

Number of Bugs

List three things you learned in this lesson.

© 2010 AIMS Education Foundation

Connecting Learning

1. What produced the sound of the Buzzin' Bug?

2. What is necessary for any sound to be made?

3. Sounds do not travel in outer space. Why do you think this is?

4. What do you think made our Buzzin' Bugs sound differently? How do you know?

5. How did our graphs compare? What caused the differences?

Connecting Learning

6. Which category had the most bugs in it? How many more did it have than the category with the fewest bugs?

7. How many of you have had a mosquito buzz in your ear? What do you think produces the buzz?

8. How would you go about finding out how many times per second it beats its wings?

9. What other things hum or buzz? Could you determine what is vibrating and its frequency? How?

10. What are you wondering now?

 © 2010 AIMS Education Foundation

Musical Instruments

Topic
Sound

Key Question
What kind of instrument can you create that makes sounds by striking, plucking, or blowing?

Learning Goals
Students will:
- learn that all sounds are created by vibrations,
- discuss how different instruments produce sound,
- use various items to create their own instruments, and
- indicate how they vibrate to make sound.

Guiding Document
NRC Standard
- *Sound is produced by vibrating objects. The pitch of the sound can be varied by changing the rate of vibration.*

Science
Physical science
   sound

Integrated Processes
Observing
Comparing and contrasting
Communicating
Classifying

Materials
Various items from home and the classroom
   (see *Management 2*)

Background Information
Sound is produced when objects vibrate. The sound thus created travels through the air in waves. For example, the plucking of a guitar string causes it to vibrate. This vibration pushes and pulls the air around the string causing disturbances in the air (sound waves) that travel outward from the string. All musical instruments cause similar waves when they vibrate. While the guitar creates sound by plucking a string, other instruments can be set into vibration by striking them or blowing into them.

Instruments that produce sound through striking are grouped into the percussion section of a band. Woodwinds make music through blowing, which sets an air column into vibration. Brass instruments, like trumpets, create sound when the player's vibrating lips set an air column into vibration. Stringed instruments, as the name implies, have strings to originate a sound.

Management
1. This activity takes advanced preparation.
2. The materials for the instruments must be collected ahead of time. Study the pages with the instrument descriptions. Decide which instruments you would like students to construct and make a list of the materials needed. Send a letter home to parents asking for help in collecting the materials on the list.
3. Make one or two instruments beforehand to show students.

Procedure
1. Have students place their hands lightly on their throats as they make a humming noise. Ask them what they feel. Explain that all sounds are produced by things that vibrate.
2. Show the students an empty can. Ask them how they might use the can to make sounds.
3. Tell them they are going to use various items to make musical instruments.
4. Brainstorm with the students how the materials could be used to make an instrument.
5. Show the students one or two instruments you have made. Have them identify the method by which each is played (blown, struck, or plucked). Play the instruments so the students can hear them.
6. Distribute the student pages. Either in groups or singly, have the students make an instrument of their choice using the instructions on the pages as a guide.
7. When the instruments are finished, have some students (out of sight of the others) play each instrument one at a time. Have the rest of the class try to guess what instrument is being played and what is vibrating to make the sound.

 © 2010 AIMS Education Foundation

Connecting Learning

1. What did you feel when you put your hand on your throat while you were humming?
2. All sounds are created by vibrations. What could you do to make a can vibrate to produce a sound?
3. To get your instrument to vibrate and produce sound, what did you have to do?
4. What vibrates in your instrument?
5. What musical instrument does your instrument look and sound like?
6. What are you wondering now?

Extensions

1. Bring several actual instruments into the classroom. Let the students feel the vibrations from various instruments as they are played.
2. Read the poem *Orchestra* by Shel Silverstein found in his book *Where the Sidewalk Ends*. Have volunteers act it out.

Musical Instruments

Key Question

What kind of instrument can you create that makes sounds by striking, plucking, or blowing?

Learning Goals

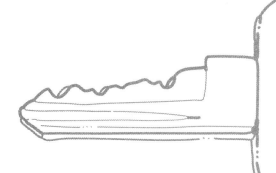

Students will:

- learn that all sounds are created by vibrations,

- discuss how different instruments produce sound,

- use various items to create their own instruments, and

- indicate how they vibrate to make sound.

Musical Instruments

You are an instrument designer. Your task is to build a musical instrument that makes sound by striking, plucking, or blowing. Use the ideas on the next two pages or come up with your own design.

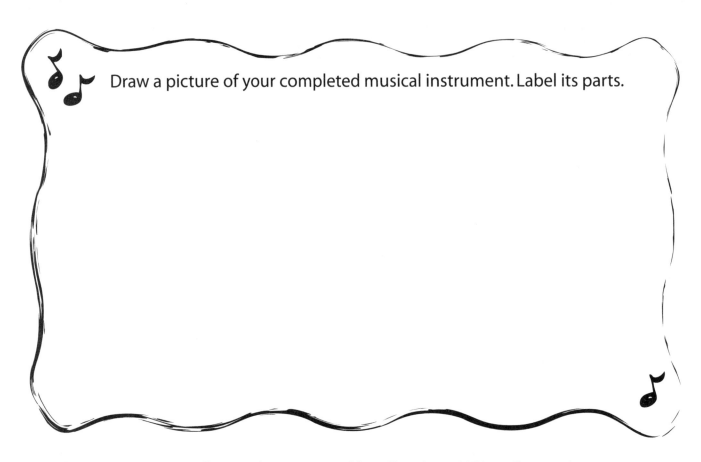

Draw a picture of your completed musical instrument. Label its parts.

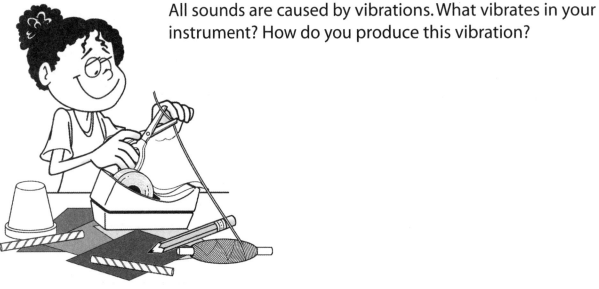

All sounds are caused by vibrations. What vibrates in your instrument? How do you produce this vibration?

© 2010 AIMS Education Foundation

Musical Instruments

Shoebox Guitar

Cut a hole in the lid of a shoebox. Stretch different width and length rubber bands around the box and pluck them. (Narrow, tight bands have a higher pitch; wide, loose bands have a lower pitch.) In this instrument, the rubber bands vibrate.

Rubber Band

Hold a rubber band tightly in your teeth and pluck the band. The sound will change as the rubber band is pulled tight or loosened. In this instrument, the rubber band vibrates. CAUTION—Don't let the band hit you in the face.

Tin Can Bass

Use a large empty metal can. Turn the can upside down and punch a hole in the bottom with a nail. Thread a heavy string through the hole and tie a paper clip to the end so it won't pull through. Tie the other end around a ruler so the string can be pulled taut. Pluck the string while varying the tension on it. In this instrument, the string vibrates.

Paper Cup Oboe

Cut the top 3 cm off a 3-oz paper cup. Cut two short slits on opposite sides of the cup, each about 1 cm long. Put a wide rubber band in the slits so that it goes across the center of the cup. Pinch the sides of the cup together so that the rubber band is held firmly in place. Blow through the small opening between your fingers and the side of the cup. Pull on the free end of the rubber band to change the pitch. In this instrument, the rubber band vibrates.

Comb Kazoo

Wrap tissue paper around a comb. Hold the kazoo up to your lips and hum gently while pursing your lips and letting them vibrate against the paper. In this instrument, the tissue paper vibrates.

Bottle Pipes

Fill several narrow-necked glass bottles with various levels of water. Blow across the opening of each bottle to produce a sound. In this instrument, the column of air in the bottle vibrates.

 © 2010 AIMS Education Foundation

Musical Instruments

Tin Can Drum

Stretch a large piece of rubber over the open end of a can. (The rubber can be cut from an old pair of rubber gloves or a large balloon.) Secure the rubber with several rubber bands. Strike the drum with the eraser end of a pencil. In this instrument, the rubber vibrates.

Spoon Chimes

Suspend various size metal spoons from strings. Strike them with a pencil. In this instrument, the metal in the spoons vibrates.

Clay Flowerpot Bell

Tie a large paper clip to the end of a piece of string. Thread the string through the hole in the bottom of a clay flowerpot. Suspend the pot by the string and strike it with a pencil. In this instrument, the clay pot vibrates.

Cardboard Box Bongos

Tape two cylindrical oatmeal boxes together. Strike them lightly with the fingertips. In this instrument, the cardboard vibrates.

Plastic Egg Maracas

Place rice, beans, or macaroni in a plastic egg and tape it shut. Shake the egg. In this instrument, the plastic egg vibrates.

Pie Tin Cymbal

Use an aluminum pie tin as a cymbal. Hold the edges of the tin with one hand and strike its bottom with the other hand. In this instrument, the metal pie tin vibrates.

Glass Bottle Xylophone

Fill several identical glass bottles with various levels of water to produce different pitches. Strike the bottles with a pencil. In this instrument, the glass bottle vibrates.

Connecting Learning

1. What did you feel when you put your hand on your throat while you were humming?

2. All sounds are created by vibrations. What could you do to make a can vibrate to produce a sound?

3. To get your instrument to vibrate and produce sound, what did you have to do?

4. What vibrates in your instrument?

5. What band instrument does your instrument look and sound like?

6. What are you wondering now?

Isn't It Interesting...
Sound Off

The songs of male humpback whales can be heard from distances of more than 30 km (20 mi) away.

0   10   20   30

Scientists believe that the blue whale makes the loudest sound of any animal. Its whistle has been measured at 188 decibels. This is a million times greater than a jet engine at about 120 decibels. (Every 10 decibels is a 10 power increase.)

Howler monkeys use their loud calls to defend and claim territory. They can be heard from 5 km (3 mi) away.

The male cicada is the loudest known insect. Its whirring sound can be heard for 400 m (1320 ft or a quarter mile).

91 © 2010 AIMS Education Foundation

We see colors because things absorb light. Red things reflect red light. They absorb the other colors. Things that are black absorb most light. They reflect very little. That is why they look black.

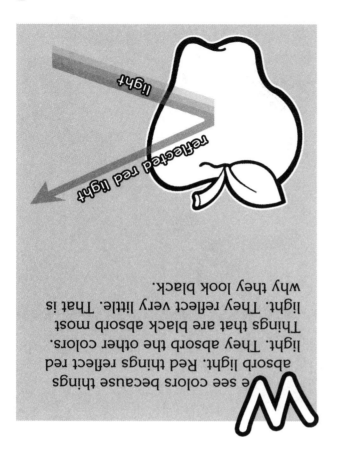

Light is a form of energy. Our primary source of light is the sun. Light energy from the sun travels through empty space and strikes the Earth.

Light can be produced in various ways. Many things that give off light also give off heat. The light from fire is due to hot, glowing particles in the flame. The sun and stars are masses of very hot gas. The light of an electric light bulb comes from tiny, hot, glowing wires.

Light Essentials

 © 2010 AIMS Education Foundation

Mirrors are a good example of things that reflect light. Most of the light that hits a mirror is reflected. This lets us see ourselves and other things in a mirror.

When light strikes an object, three things can happen.

It can be reflected.

It can be refracted.

It can be absorbed.

white light

prism

Prisms refract light. They make the light bend. This lets us see the colors that are in white light.

Light usually travels in straight lines. This is why we cannot see around corners. It is also why objects in the path of a light beam cast shadows.

© 2010 AIMS Education Foundation

FLASHLIGHT FINDINGS

Topic
Light

Key Question
How can you show that light travels in a straight line?

Learning Goals
Students will:
- observe the bright central spot of a flashlight beam and notice that this beam is visible through a tube sight, and
- use a flashlight to "draw" a straight beam of light on the floor.

Guiding Document
NRC Standard
- *Light travels in a straight line until it strikes an object. Light can be reflected by a mirror, refracted by a lens, or absorbed by the object.*

Science
Physical science
   light energy

Integrated Processes
Observing
Comparing and contrasting
Applying

Materials
For each group:
   1 flashlight
   1 toilet-paper tube
   2 rubber bands

Background Information
Light rays travel from a light source in straight lines until they encounter an object. When turned on, most flashlights have a strong central bright spot that can be seen when the light rays encounter an object. This bright spot demonstrates that light travels in a straight line, since it will move in the same direction the flashlight moves. An imaginary line extending straight out from the flashlight would be at the center of the bright spot. In this activity, a toilet-paper tube is attached to the flashlight with rubber bands and then used as a "sighting tool" to see the bright spot. Since this sighting tool is parallel to the light rays, it points to the bright spot, demonstrating that light rays travel in a straight line.

Management
1. Students should work together in small groups while doing this activity.
2. A note to parents requesting flashlights and toilet-paper tubes can supply these items. Standard flashlights using two D-cells work well for this activity as long as the batteries are fresh. Paper towel tubes cut in half will also work.
3. Students need to be cautioned not to shine the flashlights in each other's eyes.

Procedure
1. Discuss the *Key Question:* "How can you show that light travels in a straight line?"
2. Distribute the flashlights. Have students turn them on and shine them on the walls and ceiling of the classroom. Make sure students notice the bright spot at the center of the flashlight's beam that moves as they move the flashlight.
3. Give students the toilet-paper tubes, rubber bands, and the first student page. Have them attach the tube to the flashlight. Show them how to sight through the tube while the flashlight is pointing at a wall or the ceiling. They should see the bright spot through the tube, demonstrating that light travels in a straight line.
4. Show students how to turn the tube so that it is at an angle to the flashlight instead of parallel to it. (Make sure the rubber bands aren't too tight to prevent this.) Have them sight through the angled tube. They should note that the bright spot is no longer visible. This also demonstrates that the light is traveling in a straight line.
5. After a class discussion, have students complete the first student page.
6. Give students the second student page, and challenge them to use the flashlight to "draw" a straight line of light on the floor. (To do this, students will need to place the flashlight close to the floor's surface and parallel to it. This will produce a straight band of light on the surface.)
7. Have students draw a picture of how they solved this challenge.
8. End with a class discussion of the activity and how it demonstrates that light travels in a straight line.

          © 2010 AIMS Education Foundation

Connecting Learning

1. What happened to the flashlight's bright spot as you moved the flashlight? [The spot moves with the flashlight, and it always shines right where the flashlight is pointing.]

2. What did you notice when you looked through the tube? [The bright spot is visible through the tube.]

3. How does this show that light travels in a straight line? [The tube points in the same direction as the flashlight and cuts off the view of anything that is not directly in front of it. Since the bright spot is seen through the tube, the light must be traveling in a straight line.]

4. What did you do to get the flashlight to "draw" a straight line of light on the floor? [The flashlight had to be parallel to the floor and close to it.]

5. How did this show that light travels in a straight line? [The band of light on the floor goes straight out from the flashlight.]

6. What are you wondering now?

FLASHLIGHT FINDINGS

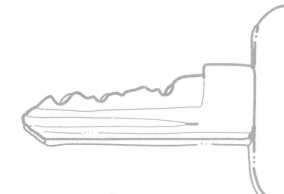

Key Question

How can you show that light travels in a straight line?

Learning Goals

Students will:

- observe the bright central spot of a flashlight beam and notice that this beam is visible through a tube sight, and
- use a flashlight to "draw" a straight beam of light on the floor.

FLASHLIGHT FINDINGS

How can you show that light travels in a straight line?

Turn on the flashlight. Shine it on the wall or ceiling. Notice the flashlight's bright spot.

Attach a cardboard tube to the flashlight with rubber bands. Look through the tube.

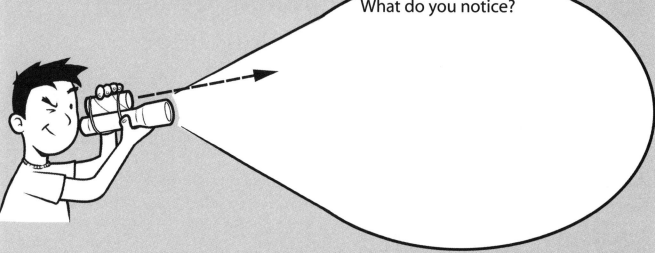

What do you notice?

Turn the tube so that it is at an angle to the flashlight. Look through the tube. Why can't you see the flashlight's bright spot?

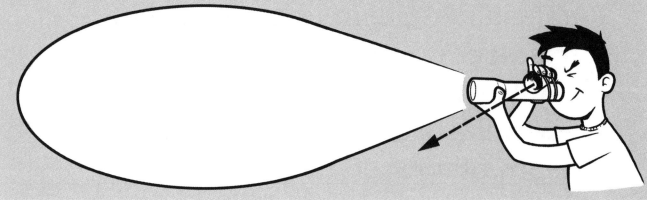

How does this show that light travels in a straight line?

 © 2010 AIMS Education Foundation

FLASHLIGHT FINDINGS

Challenge! Using only your flashlight, try to get it to make a straight band of light on the floor. Draw a picture of what you did.

 © 2010 AIMS Education Foundation

FLASHLIGHT FINDINGS

Connecting Learning

1. What happened to the flashlight's bright spot as you moved the flashlight?

2. What did you notice when you looked through the tube?

3. How does this show that light travels in a straight line?

4. What did you do to get the flashlight to "draw" a straight line of light on the floor?

5. How did this show that light travels in a straight line?

6. What are you wondering now?

 © 2010 AIMS Education Foundation

Topic
Light

Key Question
What path does light travel?

Learning Goal
Students will investigate the path light travels from a light bulb to the eye.

Guiding Document
NRC Standard
* *Light travels in a straight line until it strikes an object. Light can be reflected by a mirror, refracted by a lens, or absorbed by the object.*

Science
Physical science
    light energy
        properties

Integrated Processes
Observing
Collecting and recording data
Comparing and contrasting
Generalizing
Predicting

Materials
For each pair of students:
    D cell
    battery holder
    bulb and holder
    2 insulated wires with ends stripped, 10 cm each
    ruler
    hole punch
    scissors
    hole cards (see *Management 2*)

Background Information
Light travels along a straight-line path until it strikes an object or passes into a new medium. When light strikes an object, it can be absorbed, transmitted, reflected, or refracted. The warmth we feel from the sun is caused by the absorption of light. Light reflects off objects according to precise mathematical laws. For example, the angle at which light reflects off a flat mirror is equal to the angle at which the light hits the mirror. Light can travel through gases, liquids, and some solids. As light travels from one medium to another, it can be refracted, or bent. The refraction of light is also described by precise mathematical laws.

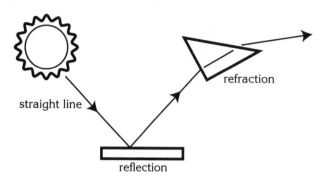

straight line

refraction

reflection

In this activity, students align the holes punched in two cards to establish the straight line between a source of light (a bulb) and the eye. They draw the straight line between the light source and the eye. Students are asked to predict where to place the third hole card so that the bulb can still be seen through all three cards. They will discover that the hole in the third card must be on the line drawn on the student page. Students add three additional cards to reinforce the observation that the bulb and the holes must lie along a line. They are then asked to think the position of 100 hypothetical cards. Students will discover that the position of any card must lie along the line from the bulb, through every hole, to the eye.

Management
1. It's important that you take the time to do this activity before presenting it to students. Doing so provides you the knowledge and skill needed to help your students maximize their learning experiences.
2. Copy the *Three Hole Cards* page on card stock. Each student will need one copy of the page.
3. Bulbs (item number 1962), bulb holders (item number 1960), battery holders (item number 1960), and wire (item number 1967) are available from AIMS.
4. Students will work in pairs. Allow time so both students can be the observers.

Procedure
1. Ask the *Key Question* and state the *Learning Goal*.
2. Distribute two *Three Hole Cards* pages, a hole punch, a ruler, and a pair of scissors to each group.

3. Instruct the students to score the page along the dashed lines and to cut each of the three cards along the solid lines. Scoring is a technique for making straight, sharp, bendable creases in card stock. To score a line, simply run a ballpoint pen or sharpened pencil over the line. Use the edge of a ruler to score neat, straight lines.

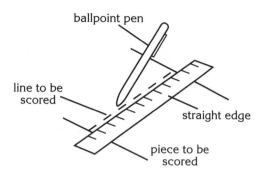

4. Tell them to fold the sides of each card along the dashed lines, fold each card in half, and, using only half the diameter of the hole punch, punch along the guide half-circle.

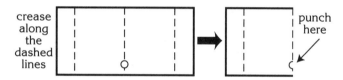

5. Distribute a D-cell, a battery holder, a bulb with bulb holder, and the connecting wires to each group. If necessary, explain how to connect the system. Tell the students to unscrew the bulb in the holder when not in use to keep from depleting the battery.

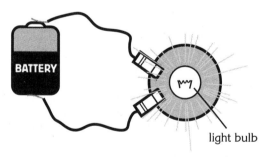

6. Distribute the student pages.
7. Instruct one student in each group to place the bulb and bulb holder at the top of the first student page. Tell those students to place the first hole card at the top of the page near the labeled position and mark the position of the hole with an X.
8. Have the students place the second card below the dashed line so that the bulb can be seen through both cards. Have the students mark the position of the second hole with an X.

9. Instruct the students to draw the path they think the light followed to pass through both holes to reach the eyes.
10. Have the same students predict the placement of the third card, between the first two cards, by placing an X on the page where the third hole will be.
11. Tell them to place the third card, turn on the bulb, and test their predictions.
12. Direct the students to borrow the three cards from their partners, and mark with an X where the three cards should be placed so that the lit bulb can be observed through the holes in all six cards.
13. Ask the students to describe the path the light from the bulb follows between the bulb and the eye.
14. Have students switch roles so both partners can be observers. Repeat the steps in the procedure.

Connecting Learning
1. What was the path you drew between the holes in the first and second card? [a straight line]
2. How did this path help you position the third card? [The hole in the card had to be on that line.]
3. This activity is meant to show that light travels in a straight line. Do you think it accomplished this or not? Explain.
4. Could the holes on the card be positioned higher or lower and still let the light pass through IF they were placed on the line? Explain your answer.
5. What can you say about the position of the hole for *any* card? [It has to be in line with the bulb, the other holes, and the eyes.]
6. How do shadows relate to the concept that light travels in a straight line? [Shadows occur when light strikes an opaque object. Because light travels in a straight line, it cannot curve around the object. The shadow is the absence of those light rays that were stopped by the opaque object.]
7. How does nighttime relate to the concept that light travels in a straight line? [Night is caused when half of the Earth has rotated out of the rays of the sun. Light from the sun strikes half of the Earth at a time; the other half of the Earth is in shadow.]
8. What are you wondering now?

HOLE CARDS

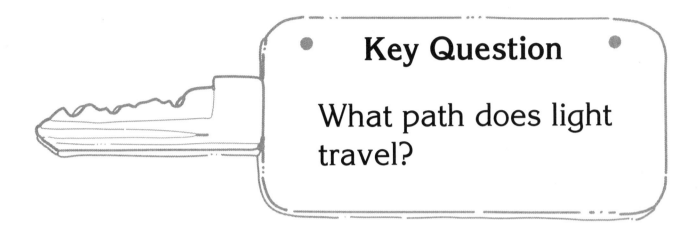

Key Question

What path does light travel?

Learning Goal

Students will:

investigate the path light travels from a light bulb to the eye.

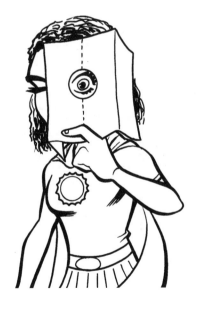

HOLE CARDS

Three Hole Cards

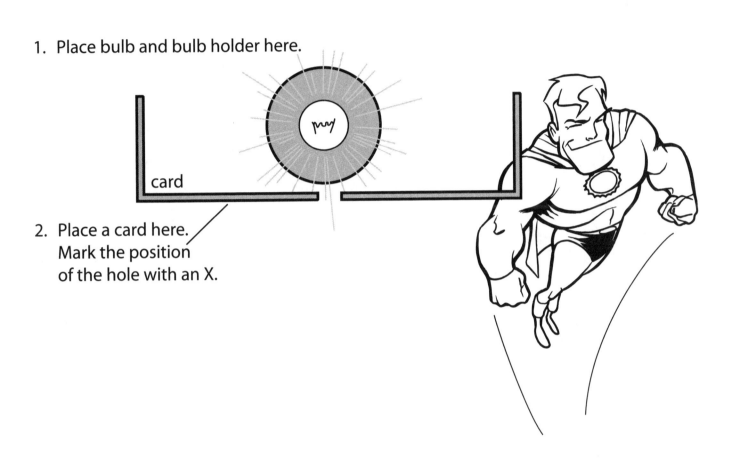

1. Place bulb and bulb holder here.

card

2. Place a card here.
   Mark the position
   of the hole with an X.

3. Place the second card below the dashed line so that the bulb can be seen through both cards.

   Mark the position of the center of the hole of both cards.

   Draw the path you think the light followed so that it passes through both hole to reach your eyes.

4. Unscrew the bulb.

   Predict the position the third card should be placed between the first two cards so that the bulb can be seen through the holes in all three cards.
   Mark the position of the hole in the third card on your page.

5. Test your prediction by screwing in the bulb to turn it on and observing whether or not the bulb is visible through all three holes.

   Describe the accuracy of your prediction.

6. Unscrew the bulb.
   Predict the positions the three cards belonging to your partner should be placed so that the bulb can be seen through the holes in all six cards. Mark the positions of the holes in the three new cards.

7. Describe the accuracy of your predictions.

8. Describe the path light follows through the air between the light bulb and your eyes.

HOLE CARDS

Connecting Learning

1. What was the path you drew between the holes in the first and second card?

2. How did this path help you position the third card?

3. This activity is meant to show that light travels in a straight line. Do you think it accomplished this or not? Explain.

4. Could the holes on the card be positioned higher or lower and still let the light pass through IF they were placed on the line? Explain your answer.

HOLE CARDS

Connecting Learning

5. What can you say about the position of the hole for *any* card?

6. How do shadows relate to the concept that light travels in a straight line?

7. How does nighttime relate to the concept that light travels in a straight line?

8. What are you wondering now?

Just Passing Through

Topic
Light

Key Question
What happens when light strikes these objects?

Learning Goal
Students will use a flashlight to discover which materials let light pass through easily, let some light pass through, or block the light.

Guiding Document
NRC Standard
- *Light travels in a straight line until it strikes an object. Light can be reflected by a mirror, refracted by a lens, or absorbed by the object.*

Science
Physical science
 light energy

Integrated Processes
Observing
Comparing and contrasting
Classifying
Communicating
Collecting and recording data
Predicting

Materials
Light source (sunshine, projector, flashlight, etc.)
A collection of items to test:
 glass jar or window glass
 sheet of white paper
 piece of clear plastic (transparency film)
 cardboard
 aluminum foil
 wax paper
 tissue paper
 glass of water
 mirror
 cloth
 book
 hand lens
 paper plate

Background Information
Light normally travels in straight lines. When light strikes an object, the object may allow light to pass through, it may block some of the light, or it may block all of the light. The object also reflects light.

Objects can be classified into three categories. A *transparent* object allows light to pass through. A *translucent* one lets some scattered light pass through but not all and will cast a light shadow. *Opaque* objects block all light and cast a dark shadow.

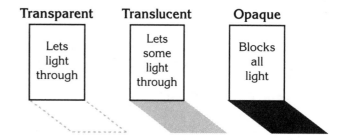

Management
1. Gather the objects you wish the students to test (see *Materials*).
2. It may be helpful to have small collections of the suggested materials in bags to be used in group settings.
3. Darken the room during testing. This makes it easier for students to decide if an object is blocking light or allowing light to pass through.
4. Mirrors (item number 1979) and hand lenses (item number 1977) are available from AIMS.

Procedure
1. Ask the students to think of all the sources of light that they can. You may want to record their suggestions on a language experience chart shaped like the sun or a giant light bulb. Sources might include a lamp, projector, match, candle, flashlight, night light, campfire, moon, glowing coals, television, fluorescent lighting, street lights, and of course, the sun.
2. Ask students if they know how light behaves. Discuss how some objects (e.g., windows) let light through very easily. Other objects (e.g., books) block all light. Then there are other objects (e.g., waxed paper) that allow some light to pass through.

3. Explain and demonstrate that
   - when an object lets light pass through easily, it doesn't cast a shadow,
   - when the light is totally blocked, the object casts a very dark shadow, and
   - when the object allows some light to pass through, it casts a dim shadow.
4. Distribute the student page and the objects you wish the students to test.
5. (This part of the lesson can be done as a whole class activity, or you can put the students in small groups to test each material with a flashlight.) Show the materials one by one. Have the students predict by a show of hands whether the object will be transparent, translucent, or opaque when a light shines on it. Then test each item by holding it up to a light source. If it casts a dark shadow, it is opaque. If the shadow is light, it is translucent. If there is little or no shadow, the object is transparent.

Connecting Learning

1. Which objects did you predict would allow light to pass through? Were your predictions correct?
2. Which objects block some light? How do those objects compare to the ones that allow light to pass through?
3. Which objects block all light? Is this what you predicted? Why or why not?
4. What evidence is there that no light passes through some objects? [a shadow is cast]
5. What is the difference between a cardboard box and a mirror? [Both objects cast a strong shadow, but the box absorbs most of the light, while the mirror reflects it.] How do you know? [You can use a mirror to redirect light; you can't do that with a box.]
6. What other objects that we didn't test would allow light to pass through? ...block some light? ...block all light?
7. What are you wondering now?

Extensions

1. Hold an opaque object close to the light source and observe the shadow. (large) Move the object away from the light source and observe the shadow. (smaller) Move the source to a different position and observe how the shape of the shadow changes.
2. Use your hands and a flashlight or overhead projector to create shadow pictures on the wall (see *Hands-on Shadows*).
3. Produce a shadow play. Put up a large sheet. The actors perform their play behind the sheet. Put a light source behind the actors so that their shadows will be projected onto the sheet. The audience seated on the other side will see only the shadows.

4. Using a bright light source, project a shadow of each student on a sheet of paper of white paper and cut out the silhouette. Trace around this silhouette on black paper, and cut out. Glue both images nose to nose for mirror reflection.

5. Make tissue paper stained glass pictures. Post in the windows for translucent beauty.
6. Play a game of shadow tag. Children are tagged by stepping on their shadows.
7. Measure shadows of children at one or two hour intervals. Observe how the length and direction of their shadows changes with the time of day.

Just Passing Through

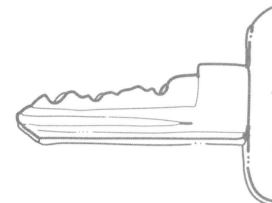

Key Question

What happens when light strikes these objects?

Learning Goal

use a flashlight to discover which materials let light pass through easily, let some light pass through, or block the light.

Just Passing Through

What happens when light hits these objects?
Will they make shadows? Test each object and mark the box.

Name of object	Lets light through	Blocks some light	Blocks all light
glass			
paper			
plastic			
cardboard			
aluminum foil			
wax paper			
tissue paper			
water			
mirror			

© 2010 AIMS Education Foundation

Just Passing Through

Hands-on Shadows

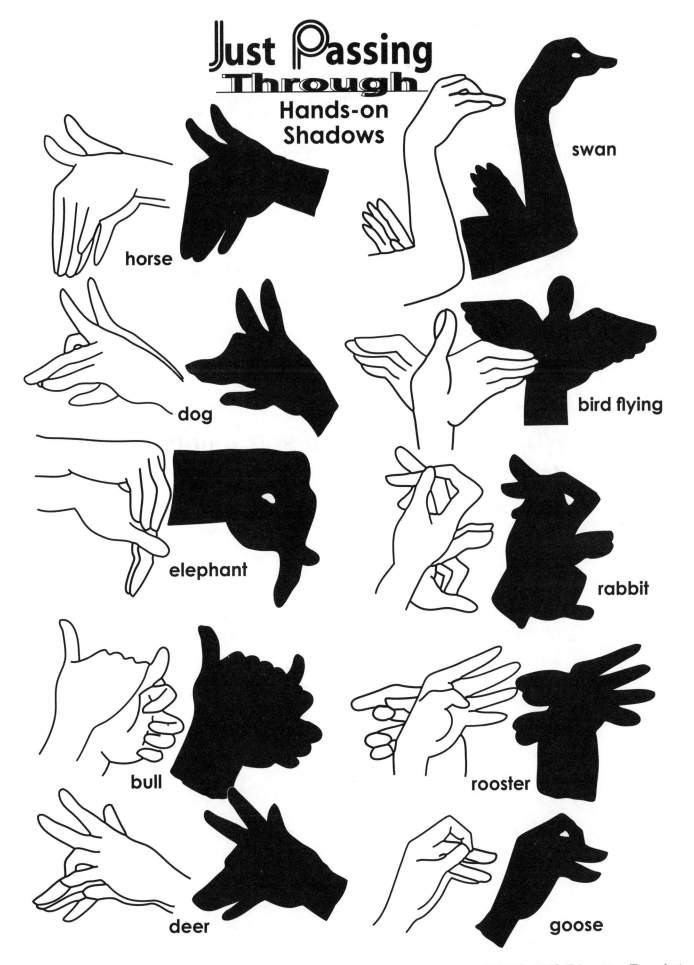

horse

swan

dog

bird flying

elephant

rabbit

bull

rooster

deer

goose

113

© 2010 AIMS Education Foundation

Just Passing Through

Connecting Learning

1. Which objects did you predict would allow light to pass through? Were your predictions correct?

2. Which objects block some light? How do those objects compare to the ones that allow light to pass through?

3. Which objects block all light? Is this what you predicted? Why or why not?

4. What evidence is there that no light passes through some objects?

5. What is the difference between a cardboard box and a mirror? How do you know?

 © 2010 AIMS Education Foundation

Just Passing Through

Connecting Learning

6. What other objects that we didn't test would allow light to pass through? …block some light? …block all light?

7. What are you wondering now?

WHAT'S BLOCKING THE LIGHT?

Topic
Light

Key Question
How does light interact with objects?

Learning Goals
Students will:
- define *transparent, translucent,* and *opaque;* and
- classify objects based on how they permit light to pass through them.

Guiding Document
NRC Standards
- *Light travels in a straight line until it strikes an object. Light can be reflected by a mirror, refracted by a lens, or absorbed by the object.*
- *Light interacts with matter by transmission (including refraction), absorption, or scattering (including reflection). To see an object, light from that object—emitted by or scattered from it—must enter the eye.*

Science
Physical science
light energy

Integrated Processes
Observing
Comparing and contrasting
Classifying
Communicating
Collecting and recording data
Predicting

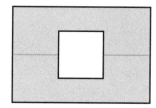

Materials
For each group:
flashlight
clear transparency film
wax paper
black construction paper
4" x 6" index cards, three per group
transparent tape

Background Information
One way in which objects can be classified is based on how light interacts with them. A *transparent* object allows light to pass through. Glass and water are examples of transparent materials. A *translucent* object permits some light to pass through it, but not all, and will cast a faint shadow. Wax paper is a good example of a translucent material. *Opaque* objects block all light and cast a dark shadow. Most objects are considered to be opaque.

Management
1. Cut a six-centimeter square piece of clear transparency film, wax paper, and black construction paper for each group. These samples will be taped over the openings in the testing frames.
2. Use the 4" x 6" index cards to make three testing frames for each group. Fold each card in half lengthwise and cut a five-centimeter square out of the center. Students will tape the squares of each material to the testing frame.

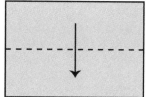

Fold index card in half.

Cut a 5-cm square out of the middle.

Completed testing frame

Procedure
1. Ask the *Key Question* and state the *Learning Goals.* If students are not familiar with the definitions of *transparent, translucent,* and *opaque,* go over them before beginning.
2. Distribute the flashlights and the other materials to each group.
3. Have students tape each of the samples to the testing frames.
4. Show students how to use the flashlight to see how each material interacts with light. Demonstrate how to hold the testing frames between the flashlight and a white piece of paper on a desk. Urge the students to find the best distance to hold the flashlight and the frames to produce the best contrast between the light and shadow.

5. Tell the students to record in words and drawings how light behaved with the material in each of the frames.
6. Challenge the students to find examples of transparent, translucent, and opaque objects in the room and record the examples in the correct sections of the student page.
7. Direct a discussion of the *Connecting Learning* questions.

Connecting Learning
1. Which objects will allow light to pass through? [transparent and translucent]
2. Which objects block all light? [opaque] What evidence is there that no light passes through some objects? [A shadow is cast.]
3. Did you find any objects in the classroom that did not let light through, but reflected most of the light? (anything shiny)
4. Why is it important to know that some materials allow light to pass through them and some materials do not? [You may want to make one room more energy-efficient and another room to have bright natural light for better reading. Window shades are constructed out of various materials to allow different levels of light to enter through windows. This way you can choose the material that best suits your needs. Another example would be in the type of glass used for bottles. Some medicines can be affected by light. Medicine bottles are often composed of materials that block light. For foods that aren't affected by exposure to light, containers are often designed to allow you to easily see their contents.]
5. If you wanted to build a shade for a window that would block out the most light from the sun, what would you use and why?
6. Based on your explorations, define *transparent, translucent,* and *opaque* in your own words.
7. What are you wondering now?

WHAT'S BLOCKING THE LIGHT?

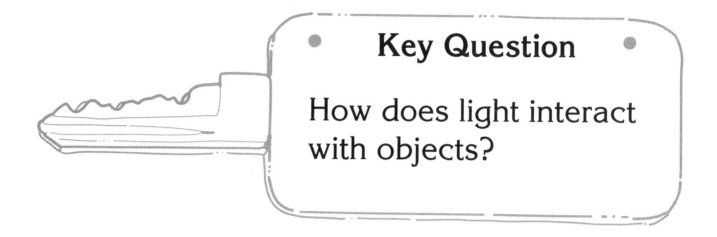

Key Question

How does light interact with objects?

Learning Goals

Students will:

- define *transparent*, *translucent*, and *opaque;* and

- classify objects based on how they transmit light (let light pass through them).

WHAT'S BLOCKING THE LIGHT?

Write about what you observe.

I'VE GOT THE BLOCKS!

Draw what you observe.

TRANSPARENT FRAME

OH BROTHER...

Write about what you observe.

Draw what you observe.

Draw what you observe.

TRANSLUCENT FRAME

Write about what you observe.

OPAQUE FRAME

 © 2010 AIMS Education Foundation

WHAT'S **BLOCKING** THE LIGHT?

TRANSPARENT MATERIALS	TRANSLUCENT MATERIALS	OPAQUE MATERIALS

Based on your explorations, define transparent, translucent, and opaque in your own words.

WHAT'S  BLOCKING THE LIGHT?

Connecting Learning

1. Which objects transmit light?

2. Which objects block all light? What evidence is there that no light passes through some objects?

3. Did you find any objects in the classroom that did not let light through, but they reflected most of the light?

4. Why is it important to know that some materials allow light to pass through them and some materials do not?

WHAT'S  BLOCKING THE LIGHT?

Connecting Learning

5. If you wanted to build a shade for a window that would block out the most light from the sun, what would you use and why?

6. Based on your explorations, define *transparent, translucent,* and *opaque* in your own words.

7. What are you wondering now?

Foiled by Oil

Topic
Light

Key Question
Would you describe a paper towel as transparent, translucent, or opaque?

Learning Goals
Students will:
- determine how light reacts on paper,
- compare how light reacts on paper with a drop of cooking oil on it, and
- use the vocabulary of *transparent, translucent,* and *opaque* to describe their observations.

Guiding Document
NRC Standards
- *Light travels in a straight line until it strikes an object. Light can be reflected by a mirror, refracted by a lens, or absorbed by the object.*
- *Light interacts with matter by transmission (including refraction), absorption, or scattering (including reflection). To see an object, light from that object—emitted by or scattered from it—must enter the eye.*

Science
Physical science
    light energy

Integrated Processes
Observing
Comparing and contrasting
Inferring
Applying

Materials
For each group:
    paper towel (see *Management 1*)
    a variety of paper (see *Management 2*)
    plastic cup, 3 oz
    cooking oil (see *Management 3*)
    sandwich size plastic bag, optional

Background Information
The *National Science Education Standards,* published by the National Research Council, says, "Light interacts with matter by transmission (including refraction), absorption, or scattering (including reflection). To see an object, light from the object—emitted by or scattered from it—must enter the eye." In this puzzling experience, students will observe light being reflected from a piece of paper. The fibers of the paper (its "hairy" nature) cause the light to be scattered in various directions. When a spot of oil is placed on the paper, the oil fills in the unevenness of the fibers and allows light to be transmitted through the paper instead of being scattered by the fibers. Students will notice that the oil smooths out the hairiness of the paper towel.

When the paper with the oil spot is on the desk, the spot looks dark; however, when the paper is held up to the light or placed over a piece of printed material, the spot seems translucent. Vocabulary of *opaque* (does not let light pass through, light is reflected and absorbed), *transparent* (can be seen through, capable of transmitting light), and *translucent* (transmits some light) should be reinforced throughout this experience.

Management
1. Brown paper towels, which are often school standard, work best for this activity.
2. Each group of three or four students needs a variety of paper besides the paper towels. Suggestions are: white copy paper, various colors of construction paper, and card stock. Cut paper into 3" x 5" pieces. Scraps of paper will also work. Putting the paper samples into small plastic bags will allow for easy distribution.
3. Each group needs a small bit of cooking oil in the plastic cup. A one-centimeter depth is sufficient.

Procedure
1. Tell students that they are going to review the terms transparent, translucent, and opaque.
2. Hold up the paper towel. Ask students which term best describes it and why they determined that. [Opaque. If we held a flashlight in front of the paper towel, the paper towel would cast a dark shadow because it blocked the light.]

3. Invite a student from each group to get a paper towel.

4. Display the *Key Question* and *Learning Goals* page. Discuss the first *Learning Goal* in relation to the paper towel. [Light did not pass through it. Light was absorbed and scattered.] Tell students that you have a variety of papers, and they are going to determine how light interacts with each one—is it transmitted, absorbed, or scattered?

5. Distribute the paper samples and let students discuss how light interacts with them.

6. Have students focus on the second *Learning Goal*. Ask them what material they need to accomplish this goal. [cooking oil] Distribute the cups of cooking oil.

7. Tell students to dip the tip of a finger into the oil and place a spot of oil on the paper towel. Discuss their observations. [The oil spot is darker than the rest of the paper.] Invite them to hold the paper up to the light and observe. [The oil spot is light, the rest of the paper is darker. The oil spot lets some light pass through. It is translucent.]

8. Direct them to hold the oil spot over some printed material and relate their observations. [The print can be read through the oil spot.] (Encourage students to use the vocabulary of opaque and translucent.)

9. Invite students to investigate different types and colors of paper, perhaps beginning with the white copy paper sample.

Connecting Learning

1. What do we mean when we say light is transmitted? [It means that light passes through a material.]

2. What are some materials that light can pass through? [Any material that is transparent or translucent, such as clear glass, transparency film, plastic water bottles, acrylic light covers, etc.]

3. Was light transmitted through our paper samples? Explain how you know.

4. Describe the surfaces of the paper. [They are fuzzy because of the paper fibers sticking up. Copy paper is smoother than the construction paper and the paper towels.]

5. What effects did the oil have on the paper? [It smoothed the surface of the paper. Some of the paper that was opaque to begin with became translucent with the oil.]

6. Do you think color had anything to do with whether a paper became translucent or not? Explain.

7. In the colonial days, instead of using expensive and fragile glass, people would smear oil on papers and place them in window frames. How does what you learned in this activity apply to this practice? [The oiled paper would somewhat protect the inhabitants, but would allow light to pass through into the home.]

8. What are you wondering now?

Extension

To make stained glass-looking pictures, have students use crayons to color a design on white paper and then use a paper towel dipped in cooking oil to coat the paper.

Foiled by Oil

Key Question

Would you describe a paper towel as transparent, translucent, or opaque?

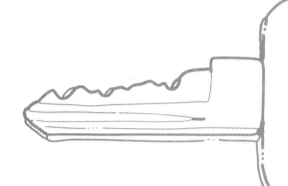

Learning Goals

Students will:

- determine how light reacts on paper,
- compare how light reacts on paper with a drop of cooking oil on it, and
- use the vocabulary of *transparent, translucent,* and *opaque* to describe their observations.

Connecting Learning

1. What do we mean when we say light is transmitted?

2. What are some materials that light can pass through?

3. Was light transmitted through our paper samples? Explain how you know.

4. Describe the surfaces of the paper.

5. What effects did the oil have on the paper?

Connecting Learning

6. Do you think color had anything to do with whether a paper became translucent or not? Explain.

7. In colonial days, instead of using expensive and fragile glass, people would smear oil on papers and place them in window frames. How does what you learned in this activity apply to this practice?

8. What are you wondering now?

 © 2010 AIMS Education Foundation

LIGHT REFLECTIONS

Topic
Light reflection

Key Questions
1. What kinds of objects reflect light?
2. What effect does the color of a reflecting surface have on the light reflected?

Learning Goals
Students will:
- explore the phenomenon of light reflection, and
- observe which colors reflect more light.

Guiding Document
NRC Standards
- *Light travels in a straight line until it strikes an object. Light can be reflected by a mirror, refracted by a lens, or absorbed by the object.*
- *Light interacts with matter by transmission (including refraction), absorption, or scattering (including reflection). To see an object, light from that object—emitted by or scattered from it—must enter the eye.*

Science
Physical science
    light energy
        reflection

Integrated Processes
Observing
Comparing and contrasting
Generalizing

Materials
Student pages
White paper, 12" x 18", for screen, one per group
Tape

Part One
For each group:
    collection of shiny objects with at least one flat surface (e.g., mirrors, CDs, pieces of aluminum foil, overhead transparencies, pieces of plastic, etc.)
    collection of non-shiny objects with at least one flat surface (e.g., books, binders, wood blocks, cardboard, paper, etc.)

Part Two
For each group:
    sheets of various colors of construction paper, including white and black

Background Information
Part One

It is a common misconception that only mirrors or similar *shiny* objects reflect light. In fact, almost *all* things reflect light. (It is quite difficult to find something that will reflect no light.) The amount of light reflected by objects varies greatly and depends on such things as an object's material, color, and texture.

Our eyes are able to see objects around us because light rays from the sun or other sources reflect off these objects and enter our eyes. If it were not for this reflected light, we could not see the objects.

In this activity, students will use objects with flat surfaces to reflect light from the sun onto a white paper screen. Although this activity uses objects with flat surfaces because they make it easier to reflect light onto the screen, it is important to note that objects without flat surfaces also reflect light.

Part Two

Sunlight is made up of all the colors of the spectrum. Things that reflect all the colors of the spectrum appear to be white in sunlight. (It is important to note that white objects only appear white when viewed in full-spectrum light. When viewed under different light, the *white* object takes on the color of the available light—the red light of a photographic darkroom for example.) Things that absorb all the colors of the spectrum appear black. Things that appear to be a certain color absorb certain parts of the spectrum and reflect other parts. This can be fairly straightforward, or rather complex. For example, a red rose appears red because it reflects red light while absorbing the other colors of the spectrum, but a yellow daffodil reflects red, yellow, and green light (which combine to produce the yellow we see) while absorbing the other colors. The light reflected from a colored (non-white or non-black) object is not full-spectrum light.

In the activity, white construction paper reflects all colors of the visible spectrum and thus should produce the brightest reflection. Black paper will absorb all colors of the spectrum and therefore should produce the dimmest reflection. The other colors will reflect various amounts of light between these two extremes. The colored papers will also produce a tint on the white screen. Students will see a red tint when using the red paper and a green tint when using the green, for example.

Management
1. **Caution students never to look directly at the sun or its reflection in a mirror (or similar shiny object) since this may result in permanent damage to their eyes.**
2. For the greatest effect, this activity needs to be done on a sunny day when there are distinct shadows. However, it can be done in the classroom.
3. You will need to think of what flat objects you can find in the classroom to use in *Part One*. If you want each group to use the same sets of objects, you will need to collect enough of each item ahead of time. For example, each group could use a CD, an overhead transparency, and a mirror for its shiny objects and a dictionary, a blue folder, and a block of wood borrowed from the kindergarten classroom for their non-shiny objects.
4. Scout out a suitable site for the activity beforehand. This site should be a place where distinct shadows and sunlit areas intersect on a smooth, flat surface.
5. If it is windy, you will need to find a way to attach the large, white-paper screen to the flat surface.

Procedure
Part One
1. Discuss the first *Key Question:* "What kinds of objects reflect light?" Use student answers to determine what students currently understand about light reflection.
2. Tell students they will be doing an activity to help them answer this question. Hand out the student page and divide the class into small groups.
3. Have students collect a variety of classroom objects (having at least one flat surface), making sure to include both shiny and non-shiny things.
4. Have students follow the directions on the student page. After they do each part, lead them in a discussion and have them record their answers.

Part Two
1. Hand out the second student page and colored construction paper.
2. Ask the second *Key Question:* "What effect does the color of a reflecting surface have on the light reflected?"
3. Have students follow the directions on the second student page. After they do each part, lead them in a discussion and have them record their answers.
4. Return to the classroom and discuss what students have learned about reflected light from doing this activity.

Connecting Learning
Part One
1. What kinds of objects reflect light? [While light is reflected better by shiny objects, non-shiny objects also reflect light.]
2. What happened when you used the shiny objects to reflect light onto the screen? [These objects reflect a bright patch of light onto the screen.]
3. What did you notice when you used the non-shiny objects to reflect light onto the screen? [These objects reflect a noticeable amount of light, but this light is more diffused than the light reflected by the shiny objects.]
4. What does this tell you about the reflection of light? [All the objects used, not just the shiny ones, reflect light.]

Part Two
1. What effect does the color of a reflecting surface have on the light reflected? [The lighter colors reflect more light than the darker colors.]
2. What happened when you used the white paper to reflect light? [The white paper reflects the most light.]
3. What happened when you used the black paper to reflect light? [The black paper also reflects light, but not as much as the other colors.]
4. What did you notice when you used the red (or another color) paper to reflect light onto the screen? [The red paper reflects red light which gives the white screen a reddish tint.]
5. What did you learn about the refection of light from this part of the activity? [All colors reflect some light, but the lighter ones reflect more than the darker ones. The color reflected corresponds to the color of the reflecting surface. The white paper seems to reflect the most light. Etc.]
6. What are you wondering now?

Extensions
1. Have students experiment with the variable of texture on the amount of light reflected.
2. Have students see what happens to the amount of light reflected when a shiny material like aluminum foil is crinkled.
3. Have students use a photographic light meter to quantify the amount of light reflected by various objects, colors, or materials.

LIGHT REFLECTIONS

Key Questions

1. What kinds of objects reflect light?

2. What effect does the color of a reflecting surface have on the light reflected?

Learning Goals

Students will:

- explore the phenomenon of light reflections, and

- observe which colors reflect more light.

DOCTOR... EVERYWHERE I GO, LIGHT REFLECTS! EXCEPT SOMETIMES... IT DOESN'T! WHAT DOES IT MEAN?

WHAT DO YOU THINK IT MEANS?

© 2010 AIMS Education Foundation

LIGHT REFLECTIONS
PART ONE

WHAT ABOUT HEAVY REFLECTIONS?

YOU HAVE GOT TO BE KIDDING ME...

What kinds of objects reflect light?

Think about this question before doing the activity. List your thoughts about this below.

Collect a number of shiny and non-shiny objects that have at least one flat surface—the more objects you can find, the better. These can be things like a mirror, a piece of aluminum foil, a piece of paper, and a book.

Take a sheet of white paper outside and place it on a vertical flat surface. This paper will be your projection screen.

One at a time, hold your objects in such a way that they reflect light onto the screen.

What happens when you do this with the shiny objects?

What happens with the non-shiny objects?

What does this tell you about the reflection of light?

 © 2010 AIMS Education Foundation

LIGHT REFLECTIONS
PART TWO

What effect does the color of a reflecting surface have on the light reflected?

Think about this question before doing the activity. List your thoughts below.

Collect several different colors of construction paper (be sure to include white and black) and use them to repeat the process you used in *Part One*.

What happens when you use the white paper to reflect light onto the screen?

What happens when you use the black paper?

What happens when you use the other colors of paper to reflect light onto the screen?

What does this part of the activity tell you about the reflection of light?

 © 2010 AIMS Education Foundation

LIGHT
REFLECTIONS

Connecting Learning

Part One

1. What kinds of objects reflect light?

2. What happened when you used the shiny objects to reflect light onto the screen?

3. What did you notice when you used the non-shiny objects to reflect light onto the screen?

4. What does this tell you about the reflection of light?

LIGHT REFLECTIONS

Connecting Learning

Part Two

1. What effect does the color of a reflecting surface have on the light reflected?

2. What happened when you used the white paper to reflect light?

3. What happened when you used the black paper to reflect light?

4. What did you notice when you used the red paper to reflect light onto the screen?

5. What did you learn about the reflection of light from this part of the activity?

6. What are you wondering now?

A Great Line Up

Topic
Mirror reflections

Key Question
What happens when you look at a line reflected in a hinged mirror?

Learning Goal
Students will use a hinged mirror to explore the reflections of a line segment.

Guiding Documents
Project 2061 Benchmarks
- *Geometric figures, number sequences, graphs, diagrams, sketches, number lines, maps, and stories can be used to represent objects, events, and processes in the real world, although such representations can never be exact in every detail.*
- *Use numerical data in describing and comparing objects and events.*
- *Mathematics is the study of many kinds of patterns, including numbers and shapes and operations on them. Sometimes patterns are studied because they help to explain how the world works or how to solve practical problems, sometimes because they are interesting in themselves.*

NRC Standards
- *Light travels in a straight line until it strikes an object. Light can be reflected by a mirror, refracted by a lens, or absorbed by the object.*
- *Light interacts with matter by transmission (including refraction), absorption, or scattering (including reflection). To see an object, light from the object—emitted by or scattered from it—must enter the eye.*

NCTM Standards 2000*
- *Recognize geometric ideas and relationships and apply them to other disciplines and to problems that arise in the classroom or in everyday life*
- *Identify, compare, and analyze attributes of two- and three-dimensional shapes and develop vocabulary to describe the attributes*

Math
Geometry
    polygons
    angles
        right
        obtuse
        acute

Science
Physical science
    light energy
        mirror reflections

Integrated Processes
Observing
Comparing and contrasting
Generalizing

Materials
Hinged mirrors
Student pages

Background Information
In this activity, students will observe what happens when a line segment is viewed through a hinged mirror. When two plane mirrors are taped together along an edge to form a hinged mirror, multiple images of an object can be seen. As the angle between the two mirrors is increased or decreased, the number of images seen also increases or decreases. For example, if the angle between the mirror surfaces is 120 degrees, two images of an object can be seen in the mirrors. For simplicity, if we call the object itself an image, then we can say we see three images. A triangle will be formed in this activity if the hinge of the mirrors is placed on the dot and the two mirrors form a 120-degree angle with the line segment as the image. Note: 3 lines at 120 degrees. If the mirrors are moved to form a 90-degree angle, four line images will make a square. (The angle of the mirrors times the number of images will always give you 360 degrees.) This mathematical relationship is not necessary for students, but is given to provide you with information. Students should generalize that the smaller the angle of the mirrors, the more sides the resulting polygons will have.

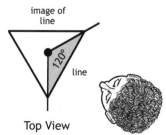

Top View

Multiple reflections of the line segment can produce both regular and irregular polygons. A polygon is a closed two-dimensional figure formed by the line segments that connect (without crossing) three or more points in a straight line. A regular polygon is a polygon in which all sides have equal length and all angles have equal measure. This lesson will focus on the regular polygons, but students will want to explore other polygons with the mirrors.

Management

1. It works best to have a hinged mirror for each student. If necessary, students can work in pairs.
2. Hinged mirrors are available from AIMS (item number 1987).
3. When students use the hinged mirrors to view the lines in this activity, make sure they move their heads down level with the mirrors and look directly into the hinged area. Caution them not to move their heads sideways because this will give them a different view of the image.

Procedure

1. Distribute the hinged mirrors to students. Allow them a few minutes to look at their own reflections in the mirrors.
2. Tell the students that the hinge of the mirror is called the vertex and that the sides of the mirrors can be moved to form angles. Invite students to hold the mirrors up and to show you a right angle. Have them show you an obtuse angle and an acute angle. Direct students to close their mirrors and put them on their desks.
3. Ask students what mirrors do. [They reflect light.] Distribute the first student page. Tell students not to use their mirrors but to try to imagine what they think they will see when they put the hinge of the mirrors on the dot and form an angle with the mirrors to view the line segment on the page.

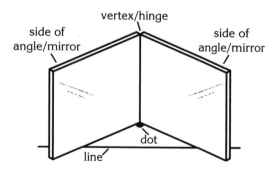

4. Have students pick up their mirrors and explore the line segment.

5. After students have had a period of exploration, ask them to put the hinge on the dot and form a right angle with the mirrors and relate what they see. [a square] Ask them to form an acute angle and relate what they see. Does the image have more sides or fewer sides than the square? Invite them to do the same with an obtuse angle.
6. Distribute the second student page. Have the students make the regular polygons with their mirrors and use a pencil to draw the angle of the mirror.
7. Have students form a generalization about the size of the angle and the number of sides the polygons have. [The greater the angle, the fewer the sides of the polygon. The smaller the angle, the more the sides of the polygon.]
8. Just for fun, let students use the mirrors to explore the line segment without the restriction of putting the hinge on the dot.

Connecting Learning

1. What do mirrors do? [reflect light]
2. If it were totally dark, would you see a reflection off a mirror? [No, there must be light present.]
3. What were some of the shapes you viewed?
4. How were the shapes alike? How were they different?
5. Where would mirrors like this be found? [in clothing stores] Why? [so customers can see different views of the clothes they are trying on]
6. What objects other than mirrors reflect light? [glass, a shiny car, the television or computer monitor when turned off, shiny metal, etc.]
7. What did you conclude about the relationship between the size of the angle and the number of images?
8. What are you wondering now?

* Reprinted with permission from *Principles and Standards for School Mathematics*, 2000 by the National Council of Teachers of Mathematics. All rights reserved.

A Great Line Up

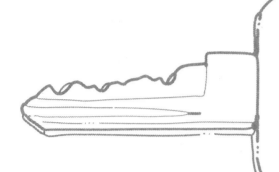

Key Question

What happens when you look at a line reflected in a hinged mirror?

Learning Goal

Students will:

use a hinged mirror to explore the reflections of a line segment.

Place the vertex of a hinged mirror on the dot. Open and close the hinged mirror.

●

What relationship do you observe between the angle of the mirror and the number of sides of the geometric figure?

A Great Line Up
A Great Line Up

Place the vertex of a hinged mirror on the dot. Adjust the mirrors to form each of the geometric figures. For each one, trace the edges of the mirrors that formed the geometric figures.

●

_____

Equilateral Triangle

●

_____

Square

●

_____

Pentagon

●

_____

Hexagon

●

_____

Octagon

●

_____

Figure of your choice

How many sides?

Name of figure:

A Great Line Up

Connecting Learning

1. What do mirrors do?

2. If it were totally dark, would you see a reflection off a mirror?

3. What were some of the shapes you viewed?

4. How were the shapes alike? How were they different?

5. Where would mirrors like this be found? Why?

6. What objects other than mirrors reflect light?

A Great Line Up

Connecting Learning

7. What did you conclude about the relationship between the size of the angle and the number of images?

8. What are you wondering now?

Make a  Kaleidoscope

Topic
Mirror reflections

Key Question
What happens when you look at objects with three mirrors hinged together?

Learning Goal
Students will use three hinged mirrors to make a kaleidoscope.

Guiding Documents
NRC Standards
- *Light travels in a straight line until it strikes an object. Light can be reflected by a mirror, refracted by a lens, or absorbed by the object.*
- *Light interacts with matter by transmission (including refraction), absorption, or scattering (including reflection). To see an object, light from the object—emitted by or scattered from it—must enter the eye.*

NCTM Standard 2000*
- *Recognize geometric ideas and relationships and apply them to other disciplines and to problems that arise in the classroom or in everyday life*

Math
Geometry
  symmetry

Science
Physical science
  light energy
    mirror reflections

Integrated Processes
Observing
Comparing and contrasting

Materials
Mylar (see *Management 1*)
Card stock
Spray adhesive
Toilet paper tubes
Brightly colored beads (see *Management 4*)
Portion cups and lids, ¾ oz
Tape
Paper towels

Background Information
In this activity, students will observe what happens when colored beads are viewed through three mirrors that are hinged together to form a triangular prism kaleidoscope. The images of the beads will be reflected in each of the three mirrors, making the image rotationally symmetric.

Management
1. Prior to doing this activity, gather and prepare the necessary materials. Kaleidoscope kits that contain Mylar mirrors, portion cups with lids, and beads are available from AIMS (item number 4135). AIMS also sells rolls of Mylar (item number 1988) and portion cups with lids (item number 4657).
2. Each student will need three Mylar mirrors that measure 1" x 3". If you plan to make your own mirrors, follow the instructions on the *Making Mylar Mirrors* page.
3. Each student will need a toilet paper tube. Paper towel tubes can be used but will need to be cut in half to make two tubes.
4. Bright multi-colored translucent tri beads work well. Tri beads can be purchased at craft stores.

Procedure
1. Ask the *Key Question* and state the *Learning Goal*.
2. Distribute the instruction sheet and materials for making the kaleidoscope.
3. Demonstrate the procedure as students follow along on the instruction sheet.
4. After students have made their kaleidoscopes, allow time for them to enjoy looking through them.

Connecting Learning
1. What makes the kaleidoscope work? [the mirrors]
2. What do mirrors do? [They reflect light.]
3. What formed the images in the kaleidoscopes? [the beads]
4. What are you wondering now?

Extensions
1. Try using small pieces of tissue paper or rice colored with food coloring or foil confetti in the portion cups.
2. Let students remove the triangular mirror system from the toilet paper tubes to view printed text, photos, etc.
3. Bring in commercial kaleidoscopes for comparison.
4. Build larger and smaller kaleidoscopes.

* Reprinted with permission from *Principles and Standards for School Mathematics*, 2000 by the National Council of Teachers of Mathematics. All rights reserved.

Make a  Kaleidoscope

Key Question

What happens when you look at objects with three mirrors hinged together?

Learning Goal

 **Students will:**

use three hinged mirrors to make a kaleidoscope.

Make-a Kaleidoscope

Making Mylar Mirrors

Materials
Card stock or poster board
Spray adhesive
Mylar
Paper
Paper cutter

Procedure

1. Cut Mylar and card stock to a manageable size such as 8" x 12". It is better to use smaller pieces than to try to work with larger pieces because it is difficult to get a smooth film of Mylar on large sections.

2. In a well-ventilated area, spray the card stock with the adhesive. CAUTION: Wisely choose the area used for spraying. The adhesive will drift and collect on nearby surfaces. Dust and dirt will then stick to the adhesive. You may want to place the card stock in the bottom of an empty paper box before spraying to help prevent adhesive drift. Old newspapers can also be used to cover nearby exposed surfaces.

3. If any large adhesive bubbles appear on the card stock, pop them before you apply the Mylar.

4. Carefully position the Mylar on the card stock. Place a sheet of paper over the Mylar surface to protect it while smoothing. Gently rub from the center of the paper to the outer edges. Lift the paper to check if the Mylar is adhered and smooth. If not, replace the paper and gently rub again.

5. If desired, draw cutting lines on the card stock side. Use a paper cutter to cut 1" x 3" mirrors. Each student will need three mirrors.

149 © 2010 AIMS Education Foundation

Make a Kaleidoscope

Materials

Three 1" x 3" pieces of card stock backed with silver Mylar
Tape
Portion cup and lid, 1 oz
Brightly colored beads
Toilet paper tube
Two paper towels

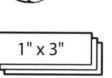

1" x 3"

Procedure

1. Join the three pieces of mirrored card stock to form a triangular prism using tape along their long edges.

    a. First, tape like this. (Leave a tiny gap between each piece.)

    b. Fold the three pieces with the mirrored surface inside and tape into a triangular prism.

2. Place a small amount of the brightly colored beads in the portion cup and secure the lid in place.

3. Position the portion cup in one end of the toilet paper tube.

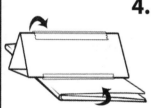

4. Fold the paper towel so that it is no longer than the mirrored prism. Wrap the mirrored prism in the paper towel and place it in the tube so that it fits snugly.

5. Point the tube toward a light source in the room or go outdoors (do not look directly at the sun) and look through the open end of the tube.

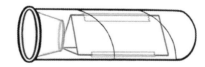

© 2010 AIMS Education Foundation

Make a Kaleidoscope

Connecting Learning

1. What makes the kaleidoscope work?

2. What do mirrors do?

3. What formed the images in the kaleidoscopes?

4. What are you wondering now?

 © 2010 AIMS Education Foundation

Light Rays SLOW DOWN

Topic
Refraction

Key Question
What happens when light travels from one transparent material to another?

Learning Goals
Students will:
- observe properties of light as it travels through different transparent materials, and
- make inferences as to why it reacts the way it does.

Guiding Document
NRC Standards
- *Light travels in a straight line until it strikes an object. Light can be reflected by a mirror, refracted by a lens, or absorbed by the object.*
- *Light interacts with matter by transmission (including refraction), absorption, or scattering (including reflection). To see an object, light from the object—emitted by or scattered from it—must enter the eye.*

Science
Physical science
   light energy
      refraction

Integrated Processes
Observing
Comparing and contrasting
Inferring
Collecting and recording data
Drawing conclusions

Materials
Clear plastic cups
Pencils
Water
Yellow crayons or sticky dots (see *Management 2*)
Student page

Background Information
When light passes through some materials such as glass or water, it looks bent. This "bending" of light as it passes from air through water is called refraction. Light travels slower through glass and water than it does through air. To bend, light must strike a surface at an angle. It does **not** bend if it goes straight in.

Management
1. Students should work in pairs. Have one student pour water into the cup while the second student observes what happens to the happy face.
2. Students may use a bright sticky dot to cover the happy face on the activity sheet or they may color it yellow.
3. Students must view the cups at about a 45° angle.

Procedure
1. Give each student or group of students a plastic cup and the student page. Tell them to color the happy face yellow or give them a bright sticky dot.
2. Demonstrate for the students how they need to view the cup at a 45° angle.

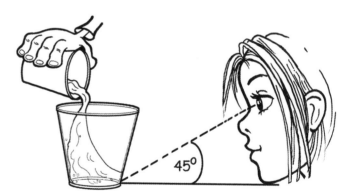

3. Direct them to slowly pour water into the cup until it is two-thirds full and observe the sticky dot. (It will look as if it is rising to the top of the cup.)
4. Have them look straight into the cup from above; the sticker is visible, but from the side—it disappears! The reflected light from the sticker is bent (refracted) and we no longer see the sticker. (You may want to place a small piece of tagboard on top of the cup so that the sticker is invisible from all angles.)

5. Next, ask the students to put a pencil inside the cup full of water. Direct them to let the pencil rest against the lip of the cup. Have them observe the pencil from above, below, and beside the cup.
6. Have the students draw what they observe.
7. For fun, let them place their thumb into the water and near the side of the cup. It will look enlarged and distorted. This is because the convex shape of the cup refracts the light and results in the magnification of the thumb.

Connecting Learning
1. What happens to the happy face when you put water in the glass? [It looks like it is rising to the top of the cup.]
2. Why can't you see the happy face on the bottom when you look at it from the side of the glass filled with water? [Light normally travels in straight lines and the reflected light from the happy face is bent (refracted) and we can no longer see it.]
3. What does a pencil look like in the cup of water? Why? [It looks broken or bent because light slows down when it travels through the glass and water, thus distorting the image.]
4. Where does the pencil appear to be broken? [at the surface of the water]
5. Why do you think it happens there? [The light's speed changes as it goes from air to water. This causes the light rays to bend (refract).]
6. Have you ever tried to catch a fish in an aquarium and found it difficult? From what you learned in this activity, why do you think it was so difficult?
7. The bending of light rays is called refraction. Besides water, what other things do you think might refract (bend) the light rays?
8. What are you wondering now?

Light Rays SLOW DOWN

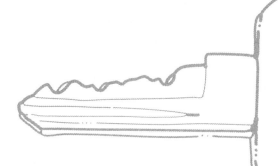

Key Question

What happens when light travels from one transparent material to another?

Learning Goals

Students will:

- observe properties of light as it travels through different transparent materials, and

- make inferences as to why it reacts the way it does.

Rays Slow Down

1. Place a pencil into a cup of water.

2. Move it around. What happens?

3. Lean the pencil against the side and draw what you see.

Light

Place cup here.

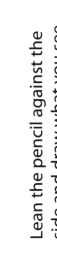

1. Color the happy face. Place the empty cup on top of the face.

2. Slowly pour water into the cup. What happens?

3. Fill the cup two-thirds full. Look through the side of the cup. What happens?

Connecting Learning

1. What happens to the happy face when you put water in the glass?

2. Why can't you see the happy face on the bottom when you look at it from the side of the glass filled with water?

3. What does a pencil look like in the cup of water? Why?

4. Where does the pencil appear to be broken?

5. Why do you think it happens there?

      © 2010 AIMS Education Foundation

Connecting Learning

6. Have you ever tried to catch a fish in an aquarium and found it difficult? From what you learned in this activity, why do you think it was so difficult?

7. The bending of light rays is called refraction. Besides water, what other things do you think might refract (bend) the light rays?

8. What are you wondering now?

RAINBOW ROUNDS

Topic
Refraction

Key Question
How can white light be changed into the colors of the rainbow?

Learning Goals
Students will:
- refract white light, and
- observe the colors that make up white light.

Guiding Document
NRC Standards
- *Light travels in a straight line until it strikes an object. Light can be reflected by a mirror, refracted by a lens, or absorbed by the object.*
- *Light interacts with matter by transmission (including refraction), absorption, or scattering (including reflection). To see an object, light from that object—emitted by or scattered from it—must enter the eye.*

Science
Physical science
   light energy
     refraction
     visible spectrum

Integrated Processes
Observing
Comparing and contrasting
Inferring
Communicating

Materials
Overhead projector
Transparent cup, 9 oz
Pitcher of water
Liter box (see *Management 2*)

Background Information
Visible light is made up of the colors red, orange, yellow, green, blue, indigo, and violet. This orderly arrangement of colors (ROY G BIV) is called a spectrum.

In this experience, students will observe the spectrum that is produced when a cup of water is placed on an overhead projector. The light is refracted by the cup and water and produces a wonderful round rainbow on the ceiling. The students will also observe the four individual spectrums on the ceiling that are produced when a liter box with water is placed on an overhead projector.

A plastic or glass triangular prism is traditionally used to separate visible light into its spectrum. Students often have trouble manipulating prisms and locating where the spectrum is projected. The experiences with the overhead projector will help them to know what they are looking for when they are given prisms.

The colors in the spectrum have different wavelengths. When white light passes through a prism, it bends. Different wavelengths bend by different amounts. Long wavelengths (red) bend less than short wavelengths (violet).

Management
1. This is a whole class observation activity. It is best done in a darkened room.
2. Liter boxes are available from AIMS (item number 1913).

Procedure
1. Turn off the lights in the classroom. Set a 9-oz transparent plastic cup on a lighted overhead projector. Invite the class to watch (observe) as you slowly begin to pour water into the cup, a centimeter at a time.
2. When the water in the cup is approximately three centimeters deep, ask the students what they see. (They will probably share what they observe projected on the screen.) Hopefully, someone will notice the rainbow on the ceiling.
3. Continue to slowly add water to the cup until the cup is about three-fourths full. Ask students to share their observations.
4. Remove the cup of water and show students the liter box. Ask them what they think will happen when you add water to the liter box.
5. Slowly add water, 100 mL at a time. When you have added approximately 300 mL, the spectrums should begin to appear. They will probably only see three; the fourth spectrum is difficult to see because it appears in the lighted part of the projection.
6. Ask students to share their observations. If you put your hand over the head lens so that light is not projected onto the screen, students will be able to see the fourth spectrum. Be careful, with some projectors, this lens is hot.

 © 2010 AIMS Education Foundation

7. Invite students to suggest other containers to try. If time allows, let students find those containers and test them.

Connecting Learning
1. What color is the light that the overhead is projecting? [white]
2. Describe the spectrum produced by the cup and water. [colorful; circular; red on the inside, blue on the outside]
3. Why do you think this happened? [The cup and the water bent the light.]
4. Describe the spectrum produced by the liter box and water. [four small colorful rainbows]
5. How are the spectrums alike? [same colors] How are they different? [in shape and size]
6. Where else do you see these spectrums? [rainbow, prism, sometimes in beveled glass, etc.]
7. What are you wondering now?

                    © 2010 AIMS Education Foundation

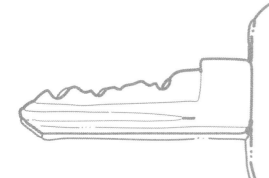

Key Question

How can white light be changed into the colors of the rainbow?

Learning Goals

Students will:

- refract white light, and
- observe the colors that make up white light.

RAINBOW ROUNDS

Sketch what you saw when you observed the cup and water on the overhead projector.

Could you distinguish separate colors? Explain.

Sketch what you saw when you observed the liter box and water on the overhead projector.

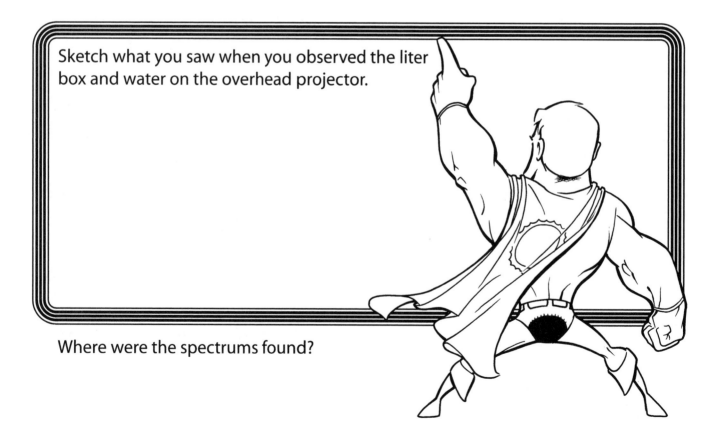

Where were the spectrums found?

          © 2010 AIMS Education Foundation

RAINBOW ROUNDS

Connecting Learning

1. What color is the light that the overhead is projecting?

2. Describe the spectrum produced by the cup and water.

3. Why do you think this happened?

4. Describe the spectrum produced by the liter box and water.

5. How are the spectrums alike?

6. Where else do you see these spectrums?

7. What are you wondering now?

© 2010 AIMS Education Foundation

CONNECTING LEARNING

Prism Play

Topic
Refraction

Key Question
What can we observe when we look at light that has passed through a prism?

Learning Goal
Students will observe the lights in their classroom and sunlight as they are refracted through a prism.

Guiding Document
NRC Standards
- *Light travels in a straight line until it strikes an object. Light can be reflected by a mirror, refracted by a lens, or absorbed by the object.*
- *Light interacts with matter by transmission (including refraction), absorption, or scattering (including reflection). To see an object, light from that object—emitted by or scattered from it—must enter the eye.*

Science
Physical science
 light energy
  refraction
   visible spectrum

Integrated Processes
Observing
Collecting and recording data
Drawing conclusions

Materials
Prisms (see *Management 3)*
White paper
Colored pencils
Student page

Background Information
 Prisms are manufactured in a wide variety of shapes. The prisms used in schools are most often triangular-shaped pieces of clear glass or plastic. One of the most common shapes is the equilateral triangle. This prism is called a *dispersion* prism because it is used to disperse light into its rainbow of colors.
 Isaac Newton was the first to use a glass dispersion prism to spread a beam of sunlight (called *white light*) into its broad *spectrum* of colors. The mnemonic ROY G BIV stands for <u>R</u>ed, <u>O</u>range, <u>Y</u>ellow, <u>G</u>reen, <u>B</u>lue, <u>I</u>ndigo, and <u>V</u>iolet—the colors of the visible spectrum.

 A beam of light that strikes any face of a prism at an angle other than 90° is bent (refracted) as it enters and leaves the prism.
 All the colors in white light travel at the same speed, but they have different amounts of energy. When a light beam strikes different media, it suddenly changes directions. This is called *refraction*. Each color has a different wavelength and is bent differently, causing the colors to separate. Red light has the longest wavelength and violet light the shortest wavelength. Red light has the lowest frequency (least energy) and violet light has the highest frequency (greatest energy) for visible light.

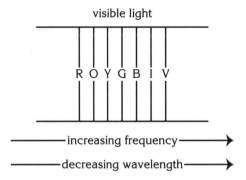

Management
1. Caution students to never look directly at the sun.
2. Students can work in groups of three or four, although each student should be given the opportunity to manipulate the prism.
3. Right angle prisms are available from AIMS (item number 1978).

Procedure
1. Ask the *Key Question* and state the *Learning Goal.*
2. Invite the students to use the prism as an eyepiece and look at the classroom lights. Encourage them to rotate the prism in all ways while viewing the lights.
3. Discuss their observations.
4. Inform the students that they will be going outside to observe what happens when sunlight passes through the prism. Tell them they will try to focus the spectrum on a sheet of white paper (or on a concrete slab if available). Warn them to not look directly at the sun!
5. Go outside and let students manipulate the prisms until they see a spectrum. Have them note the order of the colors.

6. Go back inside the classroom. Distribute the student page and have students write the order of the colors and then use their colored pencils to finish the page.

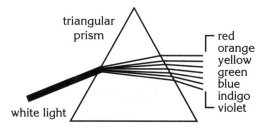

triangular prism

white light

red
orange
yellow
green
blue
indigo
violet

Connecting Learning

1. What happens to light that enters a prism? [It is refracted (bent). With the prisms we used, it was also separated into the colors of the visible spectrum.]
2. Describe what you observed when you looked through the prism at the lights in the classroom.
3. What did you observe when you looked at the spectrum from the sunlight?
4. Was there a difference between what you saw when you looked at the electric lights and the spectrum from the sunlight? Explain.
5. What conclusion can you draw about what makes up sunlight?
6. What are you wondering now?

 © 2010 AIMS Education Foundation

Prism Play

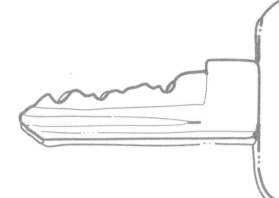

Key Question

What can we observe when we look at light that has passed through a prism?

Learning Goal

Students will:

observe the lights in their classroom and sunlight as they are refracted through a prism.

© 2010 AIMS Education Foundation

Prism Play

Color and label to match what you observe.

White light is made of colors mixed together. A prism refracts (bends) light and shows us the colors it contains.

© 2010 AIMS Education Foundation

Prism Play

Connecting Learning

1. What happens to light that enters a prism?

2. Describe what you observed when you looked through the prism at the lights in the classroom.

3. What did you observe when you looked at the spectrum from the sunlight?

4. Was there a difference between what you saw when you looked at the sunlight and the electric lights and the spectrum from the sunlight? Explain.

5. What conclusion can you draw about what makes up sunlight?

6. What are you wondering now?

          © 2010 AIMS Education Foundation

Visible light is made up of the colors of the rainbow starting with red. Each one gradually blends into the next color. When all of the rainbow colors mix together, we get white light.

2

Light is a form of energy. It travels in waves. Each color in the spectrum has a different wavelength. Our eyes can only see a limited range of light waves. The range that we can see is called the *visible spectrum.*

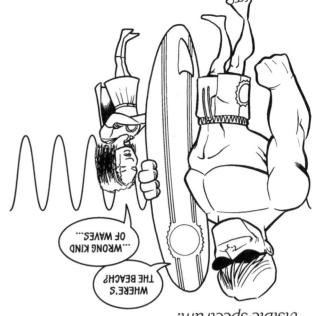

Most people can see the colors in the visible light spectrum. However, the visible light spectrum is not the same for all creatures. For example, bees can see ultraviolet light, but they can't see the color red that we see. Crocodiles miss out on a lot because they can only see black, white, and shades of gray.

The Electromagnetic Spectrum

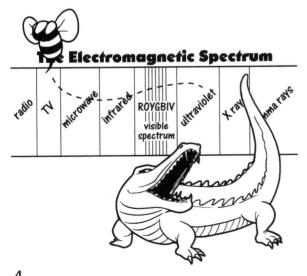

The Lowdown on Light

A green leaf reflects green more strongly than the other colors, so we only notice the green.

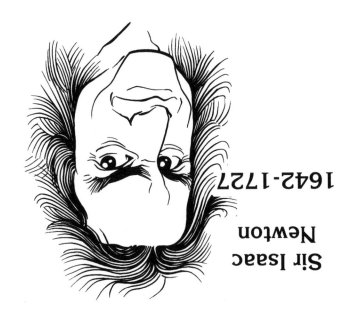

Sir Isaac Newton

1642-1727

Around the year 1700, Sir Isaac Newton did many experiments with light. He was the first to understand that white light is a combination of all the colors of the rainbow and every shade in between.

White is the mixture of all colors. You can observe the colors that make up white light by placing a prism in light's path. The prism bends (refracts) the light, separating it into the visible color spectrum.

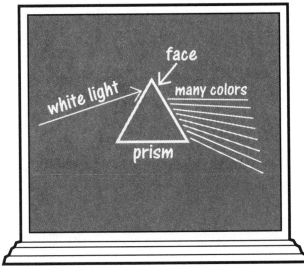

white light → face → many colors
prism

AROUND THE BEND

© 2010 AIMS Education Foundation

Newton studied the different colors. He put a screen with a hole in it behind the wedge. This way he could let only one color at a time pass through. He observed colored objects in the separate beams of colors.

The light was white as it entered the glass. He observed different colors in the light that came through the glass.

Newton concluded that an apple looks red to us because it reflects red light better than others; the other colors are absorbed by a red apple.

Newton sat in a dark room where a ray of sunshine came in through a small hole in a curtain. He then put a small wedge of glass into the path of the light beam.

3

Convex Lens

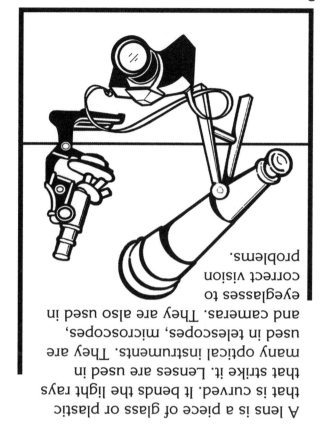

light

Some lenses bulge outward. They are thicker in the middle than on the outside. These are called **convex lenses**. They are used in magnifying glasses. Some people may have trouble seeing things up close. They can get eyeglasses with convex lenses that will help them see more clearly.

A lens is a piece of glass or plastic that is curved. It bends the light rays that strike it. Lenses are used in many optical instruments. They are used in telescopes, microscopes, and cameras. They are also used in eyeglasses to correct vision problems.

Some lenses are thinner in the middle. They cave inward. These are called **concave lenses**. People who have trouble seeing things that are far away can be helped with eyeglasses that have concave lenses.

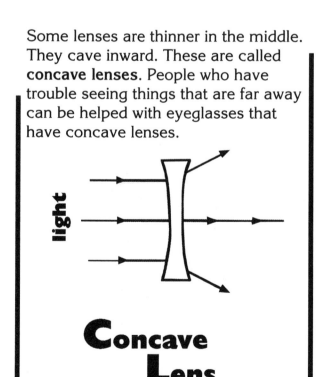

light

Concave Lens

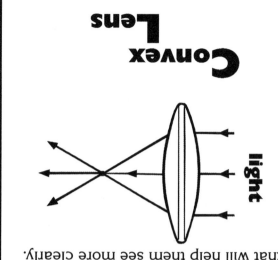

Lenses

Homemade Microscopes

Topic
Microscopes

Key Question
How can you make your own simple microscope?

Learning Goals
Students will:
- learn about Antony van Leeuwenhoek and his discoveries using simple microscopes,
- create their own simple microscopes using clear plastic cups or bottles filled with water, and
- make careful observations about an item of their choice.

Guiding Documents
Project 2061 Benchmarks
- *Throughout all of history, people everywhere have invented and used tools. Most tools of today are different from those of the past but many are modifications of very ancient tools.*
- *Technology enables scientists and others to observe things that are too small or too far away to be seen without them and to study the motion of objects that are moving very rapidly or are hardly moving at all.*
- *Microscopes make it possible to see that living things are made mostly of cells. Some organisms are made of a collection of similar cells that benefit from cooperating. Some organisms' cells vary greatly in appearance and perform very different roles in the organism.*

NRC Standards
- *Light travels in a straight line until it strikes an object. Light can be reflected by a mirror, refracted by a lens, or absorbed by the object.*
- *Simple instruments, such as magnifiers, thermometers, and rulers, provide more information than scientists obtain using only their senses.*
- *Employ simple equipment and tools to gather data and extend the senses.*

- *Tools help scientists make better observations, measurements, and equipment for investigations. They help scientists see, measure, and do things that they could not otherwise see, measure, and do.*
- *Science and technology have been practiced by people for a long time.*

Science
Physical science
  light energy
    convex lenses
    magnification
Technology
  microscopes

Integrated Processes
Observing
Collecting and recording data

Materials
Antony van Leeuwenhoek and His Microscopes rubber band book
Rubber bands, #19
Clear plastic containers (see *Management 1*)
Items for observation (see *Management 2*)
Student pages

Background Information
Antony van Leeuwenhoek (An toe´ nee van Lay´ wen-hook) was born in 1632 in a small village in the Netherlands. As an adult, he owned a draper and haberdashery shop, and it was as a result of this that he came to make the lenses that he eventually used to make many important discoveries in microbiology. In seeking to create a more effective glass for inspecting cloth, he created a simple microscope that he used to look at muscle tissue, hair, ivory, and the parts of a bee. Intrigued, he began to spend his spare time grinding lenses and observing small specimens from plants and animals. Throughout his life he made more than 400 lenses, some of which magnified up to 270 times.

Although van Leeuwenhoek had no formal scientific training, he had a keen intellect, excellent vision, great manual dexterity, and mathematical precision. By combining these with careful observations and detailed descriptions, he was able to make many groundbreaking discoveries. He was the first person to observe bacteria, and was also the first to accurately describe red blood cells. He disproved the theory of spontaneous generation, and described the role that the egg and sperm play in fertilization.

All of van Leeuwenhoek's discoveries were made with simple microscopes that used only one lens. Any clear object with a curved surface will magnify to some extent. Simple magnifiers can be made using plastic cups or bottles full of water. With careful observation, students can make their own discoveries using simple microscopes.

Management

1. Collect enough clear plastic cups or bottles for each group of students to have one. Bottles are easier to use because they can be capped and laid on their sides; however, you must find bottles that have smooth sides and remove any labels ahead of time. Plastic cups work, but are more difficult to use because whatever is being observed must be held behind the cup.

2. Have a variety of items for students to observe using their homemade microscopes. These could include dead insects, parts of plants or flowers, rocks, pieces of fruits or vegetables, etc. You can also allow students to collect their own items from the playground or classroom.

Procedure

1. Distribute the rubber band book on Antony van Leeuwenhoek. Read through it as a class and discuss his contributions to science.

2. Explain that Antony used simple microscopes that had only one lens. They were more like a modern magnifying glass, but much more powerful. The microscopes that scientists use today have multiple lenses.

3. Point out that even though Antony had no formal scientific training, he made many very important discoveries because he made careful observations and described what he saw accurately.

4. Tell students that they will be given the chance to make their own simple microscopes and discover some things they may not have known by observing objects up close.

5. Divide students into groups and distribute the materials for the microscopes. Explain to students that they will need to completely fill the bottle (or cup) with water before it will magnify.

6. Allow students to select items from your collection or to find their own items to observe.

7. Distribute the student page and give time for students to make their observations and record what they discover.

8. Have the groups share their discoveries with the class.

Connecting Learning

1. How is your homemade microscope like a real microscope? [It magnifies objects.] How is it different? [It only has one lens; it is filled with water; etc.]

2. What type of surface do you need in order to magnify objects? [a (convex) curved surface that is transparent]

3. What specimen did your group select? What things could you see when you observed it without a microscope?

4. What did you discover when you observed your specimen under your homemade microscope?

5. What kinds of things might you see if you had a more powerful microscope?

6. What other things could you use to make a homemade microscope?

7. What are you wondering now?

Extensions

1. Have groups look at the same specimens using hand lenses and microscopes and compare the amount of detail they are able to see.

2. Experiment with ways to make the homemade microscopes more powerful. Does the size of the bottle make a difference? Is glass more effective than plastic?

3. Find additional objects that can also be used to magnify and compare the amounts of magnification. [fishbowl, glass jar full of water, etc.]

 © 2010 AIMS Education Foundation

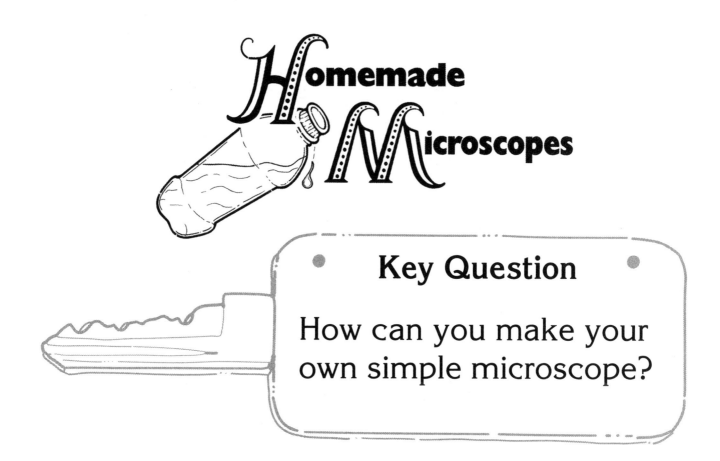

Homemade Microscopes

Learning Goals

Students will:

- learn about Antony van Leeuwenhoek and his discoveries using simple microscopes,
- create their own simple microscopes using clear plastic cups or bottles filled with water, and
- make careful observations about an item of their choice.

1

Antony van Leeuwenhoek and His Microscopes

8

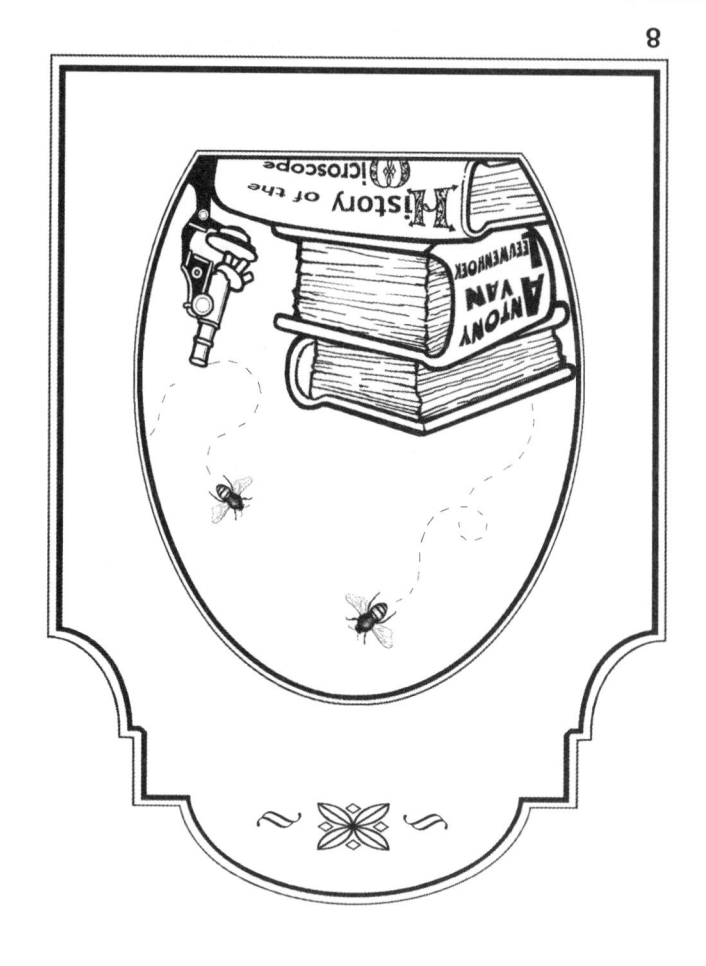

2

Antony van Leeuwenhoek was born in 1632 in Delft, Netherlands. He owned a shop that sold cloth. One day, he wanted to improve the glasses used to inspect cloth. He put a small ball of glass between metal plates. This lens worked like a simple microscope.

7

He always made careful observations. He described what he saw accurately. This made the use of the microscope more popular. Antony van Leeuwenhoek made many discoveries that are very important to science.

He also studied animals. People used to believe that some animals came from mud or sand. Antony proved that this was not true. He even carried flea eggs in his pocket so he could see what happened when they hatched.

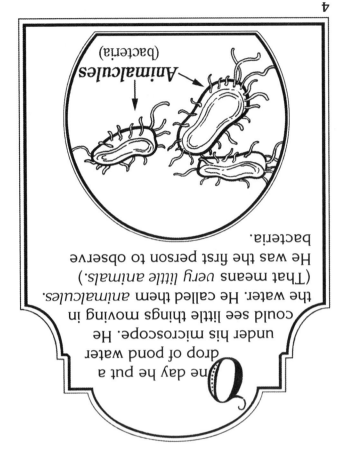

Animalcules (bacteria)

One day he put a drop of pond water under his microscope. He could see little things moving in the water. He called them animalcules. (That means *very little animals*.) He was the first person to observe bacteria.

Throughout his life, Antony kept making lenses. He made more than 400 total. Some were as small as the head of a pin. They got more and more powerful. Some magnified up to 270 times.

He started to look at things besides cloth through this lens. He looked at hair, ivory, and parts of a bee. Soon, Antony began to make more lenses. He used these to study many things.

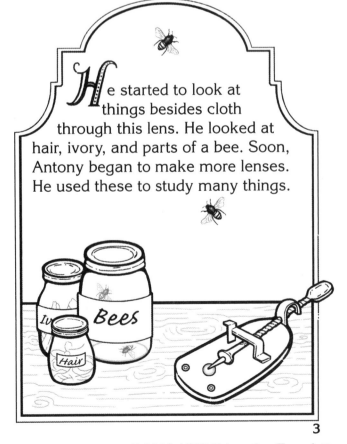

Bees

Hair

Iv

© 2010 AIMS Education Foundation

Homemade Microscopes

Fill your bottle with water. Cap the bottle. Select your item to observe. Draw and describe your item here.

Place the bottle on its side over the item. Adjust the height of the bottle until the item is in focus. Draw and describe what you see in the magnified view.

182 © 2010 AIMS Education Foundation

Connecting Learning

1. How is your homemade microscope like a real microscope? How is it different?

2. What type of surface do you need in order to magnify objects?

3. What specimen did your group select? What things could you see when you observed it without a microscope?

4. What did you discover when you observed your specimen under your homemade microscope?

5. What kinds of things might you see if you had a more powerful microscope?

6. What other things could you use to make a homemade microscope?

7. What are you wondering now?

Topic
Microscopes

Key Question
What interesting things on a penny can you observe using the magniviewer?

Learning Goal
Students will construct and use a microscope made with a hand lens and a half-pint milk carton.

Guiding Documents
Project 2061 Benchmark
* *Technology enables scientists and others to observe things that are too small or too far away to be seen without them and to study the motion of objects that are moving very rapidly or are hardly moving at all.*

NRC Standards
* *Light travels in a straight line until it strikes an object. Light can be reflected by a mirror, refracted by a lens, or absorbed by the object.*
* *Simple instruments, such as magnifiers, thermometers, and rulers, provide more information than scientists obtain using only their senses.*
* *Tools help scientists make better observations, measurements, and equipment for investigations. They help scientists see, measure, and do things that they could not otherwise see, measure, and do.*

Science
Physical science
    light energy
        convex lenses
        magnification

Integrated Processes
Observing
Comparing and contrasting
Recording data

Materials
For each microscope:
    half-pint milk carton
    mini hand lens (see *Management 1*)
    scissors
    metric ruler
    masking tape
    shiny penny

Background Information
A convex lens is thicker in the middle than it is on the sides. It can be used to magnify objects. The magnifying power of a single lens is defined by the number of times an object viewed through the lens is magnified. For example, if the object viewed through the single lens is six times larger than the actual object, the lens is described as a six power lens (6X).

To begin this activity, students will use a convex lens in the construction of a simple microscope to enhance their observational skills as they view the front and back sides of a penny.

A one-centimeter slice of the base of a half-pint milk carton is cut off to be used as a stage for the magniviewer. The opening that results from cutting off the base is then positioned toward a window or other bright area in the classroom to allow light to enter the magniviewer for better viewing of the object. To focus the magniviewer, students will raise and lower the stage inside the milk carton. This will be especially important if they are using more than one lens (see *Extensions*).

When viewing the shiny penny through the magniviewer, students may notice a letter under the minting date. The presence or the absence of the letter identifies where the penny was minted. Philadelphia and Denver are the two locations where pennies are made. The letter *D* specifies the Denver Mint. If no letter exists under the minting date, the penny was minted in Philadelphia. Three minute initials, *VDB*, are found at the base of Lincoln's right shoulder. These are the initials of Victor David Brenner, the sculptor who was singled out by President Theodore Roosevelt for formulating the design of the Lincoln bust. On the tail side of the penny, students may notice the figure of Lincoln sitting in a chair in the Lincoln Memorial. They may also notice the letters *FG* on the right side of the building near some shrubbery. These letters stand for Frank Gasparro who was the Assistant Engraver at the Philadelphia Mint in 1959. To mark the 150th anniversary celebration of Lincoln's birth, Gasparro prepared the winning entry of the Lincoln Memorial that was to change the tail side of the penny, which had been wheat ears. Students may also observe the Latin phrase *E PLURIBUS UNUM* that translates "one out of many" in reference to one government formed out of many states. The observations of these unusual things may prompt students to research their meanings. Encourage their questions and guide them in their search for resources. Students tend to learn more when they pursue their own questions and interests and report their findings rather than to have information dispensed by the teacher.

Management

1. Small, high quality hand lenses called mini hand lenses (item number 3072) are available from AIMS.
2. Prior to doing this activity, collect half-pint milk or juice cartons. Wash them and let them dry. Each student will need one carton.
3. If students are viewing light-colored objects or materials with their magniviewers, have them put a piece of black construction paper on the stage to provide a greater background contrast.
4. Students should be allowed a time of free exploration using their magniviewers. During this time they will learn about adjusting the focus by raising and lowering the stage and about directing the opening of the milk carton toward the light.

Procedure

1. Distribute pennies for students to observe. After a period of time, collect them and have students draw their recollections of the head side, tail side, or both sides. For making their illustrations, they can use scratch paper.
2. Tell the students that they are going to make a simple microscope to observe the pennies.
3. Ask the *Key Question* and state the *Learning Goal.*
4. Distribute the directions for constructing the microscopes, the milk cartons, scissors, tape, and mini hand lenses.
5. Go through the steps one by one, assisting as needed.
6. Allow a period of free exploration.
7. Discuss what students learned during their free exploration.
8. Distribute the student recording page. Ask students to look at the head side of a shiny penny. (Rolled masking tape will secure the penny to the stage.) Have them draw what they see. One technique for helping students to illustrate details is to have them look a little, draw a little, look a little, draw a little, etc. If they don't notice the three letters at the base of Lincoln's right shoulder, ask them to look in that area. Once they have located the letters, ask them what they think the letters might mean.
9. Have students draw the tail side of the penny. Direct them to draw what they see.
10. Have students look at the first illustrations of the penny and compare them to their illustrations done with the simple microscope.
11. Encourage students to do some research to find the meanings of the letters under Lincoln's shoulder, the letter under the year of minting (if one exists), and other interesting aspects of the penny's design.

Connecting Learning

1. What things were you able to see on the penny when you used the magniviewer that you hadn't noticed before?
2. Why were you able to see these things? [They were magnified.]
3. What things do you want to know about the penny? How could you find the answers?
4. For what other things can you use this magniviewer?
5. When you rub your fingers over the surface of the magnifying lens, what do you notice? [It's not flat, it's curved.] Why do you think it's curved? What other transparent objects can you think of that have a curved surface? [clear film containers, medicine vials, jars, etc.] What might you do to get these objects to magnify like a hand lens? [add water] Design a plan to test your idea.
6. How would lenses that magnify help scientists?
7. Why do you think this microscope is called a magniviewer?
8. What are you wondering now?

Extensions

1. Add another magnifying lens to the inside of the milk carton to double the power of the microscope. Ask students what they need to do in order to bring the viewed object into focus. [raise it] Also ask whether they can now see a greater or lesser part of the object. [less because what they see is now a larger image than what they saw before]
2. Make mystery-scopes by attaching an object to the stage of the microscope with rolled masking tape. Then seal the open end of the milk carton by taping waxed paper to it. Ask students to identify the object that they observe. Use objects or materials such as a raisin, pencil shavings, hook and loop material (i.e., Velcro), burdock, steel wool, soap pad, etc.
3. Cut a hole in the center of the stage and one in the milk carton opposite the hole in the stage (on the bottom of the magniviewer carton). Have students lift the magniviewer up from the tabletop so that light will be able to pass up through the holes and through translucent specimens on the stage, just like with a compound microscope. This allows one to view prepared slides.

Home Link

Have students bring objects or materials from home to put in the mystery-scopes.

Magniviewer

Key Question

What interesting things on a penny can you observe using the magniviewer?

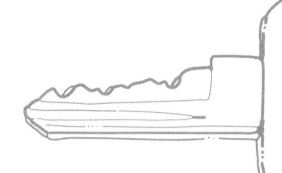

Learning Goal

Students will:

construct and use a microscope made with a hand lens and a half-pint milk carton.

Magniviewer

Construction Instructions

1.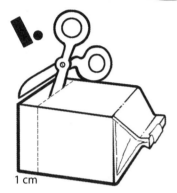
Tape the top of the milk carton closed. Cut off the bottom of the milk carton and save it for later.

1 cm

2.
Draw an X on the side of the carton.

3.
Trace around a penny in the middle of the X.

4.
Cut down the middle of the carton and around the circle that you traced.

5.
Tape the magnifying lens over the hole of the milk carton.

6.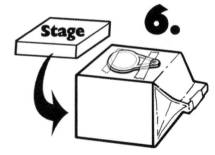
Stage
Put the cut-off section (first step) inside the milk carton. This is your microscope stage.

7.
Turn the open end of your microscope toward the window or light for brighter viewing. Slide the stage up or down for focusing.

 © 2010 AIMS Education Foundation

Magniviewer

Draw what you see when you look at a penny through the magniviewer.

Heads

Tails

What things on the penny were you able to observe with the magniviewer that you didn't see without it?

Compare your observations with those of your classmates. Look at similarities and differences. What explanation do you have for the differences?

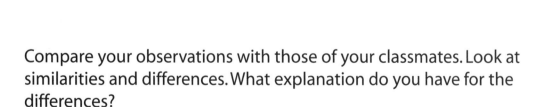

          © 2010 AIMS Education Foundation

Connecting Learning

1. What things were you able to see on the penny when you used the magniviewer that you hadn't noticed before?

2. Why were you able to see these things?

3. What things do you want to know about the penny? How could you find the answers?

4. For what other things can you use this magniviewer?

          © 2010 AIMS Education Foundation

Connecting Learning

5. When you rub your fingers over the surface of the magnifying lens, what do you notice? Why do you think it's curved? What other transparent objects can you think of that have a curved surface? What might you do to get these objects to magnify like a hand lens? Design a plan to test your idea.

6. How would lenses that magnify help scientists?

7. Why do you think this microscope is called a magniviewer?

8. What are you wondering now?

 © 2010 AIMS Education Foundation

Isn't It Interesting...
Looking at Lenses

The oldest known lens was made of polished rock crystal. It was found in the ruins of ancient Nineveh. Nineveh was located in what is now northern Iraq, along the Tigris River.

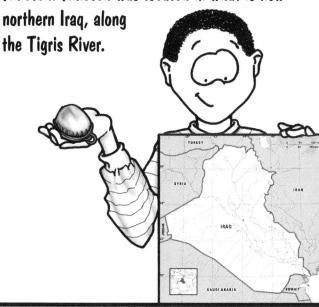

Aristophanes, a Greek playwright who lived between 450-388 BC, wrote about using glass for erasing writing from wax tables.

It's been claimed that Seneca, a Roman rhetorician who was born about 4 BC, read "all the books in Rome" by using a glass globe of water to magnify the text.

Reading stones, magnifying glasses as we know them today, were developed around 1000 AD. They were used by farsighted monks as an aid for reading.

 © 2010 AIMS Education Foundation

Nuclear energy can produce a lot of heat. In a reactor, water is heated by nuclear energy. The hot water turns to steam. The steam is used to make electricity.

All living things use heat energy. Plants and animals use energy from the sun. Humans need heat to cook foods, provide light, power machines, run automobiles, dry clothes, and provide comfort.

The Earth contains heat deep inside. Some of that heat escapes through geysers. This heat can be used to make electricity.

8

ENERGY EXPLORATIONS

© 2010 AIMS Education Foundation

Fire is a source of heat. Wood, natural gas, and oil produce heat when burning.

4

The sun is our most important source of heat. The light energy from the sun strikes matter on Earth. Some of this energy is changed to heat. Our planet would be too cold for us if it were not for the heat from the sun.

Anything that gives off heat is a source of heat. The heat we use on Earth comes from several sources.

Heat is produced when one object rubs against another. This is called friction. Friction helps us stop our roller blades. It keeps our shoes from sliding on the floor. Friction is often an unwanted source of heat because it may damage objects.

Solar Mitts

Topic
Heat energy

Key Question
Which colors feel warmest in the sun?

Learning Goal
Students will use their sense of touch to compare how different colors absorb the sun's heat energy.

Guiding Documents
Project 2061 Benchmark
- *When warmer things are put with cooler ones, the warm ones lose heat and the cool ones gain it until they are all at the same temperature. A warmer object can warm a cooler one by contact or at a distance.*

NRC Standards
- *The sun provides the light and heat necessary to maintain the temperature of the earth.*
- *Heat can be produced in many ways, such as burning, rubbing, or mixing one substance with another. Heat can move from one object to another by conduction.*
- *Heat moves in predictable ways, flowing from warmer objects to cooler ones, until both reach the same temperature.*

Science
Physical science
    heat energy
        radiation
        temperature

Integrated Processes
Observing
Comparing and contrasting
Drawing conclusions

Materials
Five colors of 9" x 12" construction paper
    (see *Management 1* and *2*)
Tape or staplers
Crayons or colored pencils

Background Information
The sun is the major source of heat energy on Earth. The sun's rays travel through space, striking everything on Earth exposed to them; heat is being transferred by *radiation*. When you stand in the sun, you feel the sun's warmth. In the shade, you feel cooler because you are shielded from the sun's rays.

Surfaces vary in how much of the sun's heat energy they reflect or absorb, partly due to color. A white surface appears white because it reflects most of the sun's light (and heat) back into space. A black surface appears black because it reflects very little light and heat. Since black surfaces absorb more heat, they become warmer. White and black are at the two heat absorption extremes, with other colors falling somewhere between them on a continuum. In general, darker colors will be closer to black and lighter colors closer to white.

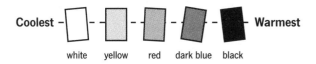

Coolest — white — yellow — red — dark blue — black — Warmest

Holding their covered hands out to the sun, students will use their sense of touch to feel differences in the heat energy collected by the colored mitts. The difference between black and white is dramatic. Other colors may take several comparisons to order. This qualitative experience can easily be related to real life—how the colors of clothing, cars, and buildings affect our comfort.

Management
1. Black and white paper are mandatory for this investigation. It is suggested that the three primary colors—yellow, blue, and red—be used to complete the palette; however, other colors can be substituted or added if you wish.
2. To control variables, construction paper should be of the same weight and texture. White paper is frequently different than the others.
3. Organize the class into groups of three or four.
4. This activity can be done at any time of the year as long as sunshine is not interrupted by too many clouds.

Procedure

1. Ask the *Key Question:* "Which colors feel warmest in the sun?"
2. Give groups the activity page and materials. Have them follow the directions for making the mitts.
3. Take the class outside to a sunny location. To better feel the sun's energy on the hands and not warm up the whole body, it is preferable for students to stand in the shade and hold just their mitt-covered hands in the sun.
4. Have groups follow directions for using the mitts, resulting in the colors being ranked from coolest to warmest.
5. Discuss the investigation.

Connecting Learning

1. What was the source of heat for this activity? [the sun]
2. Did the mitts of all colors feel the same? Explain.
3. Which colors were easy to place? [most likely black and white]
4. Which colors were harder to put into order? How did your group solve any disagreements?
5. How do our class results compare? (Groups may or may not be in complete agreement with each other, a good opportunity for further discussion.)
6. How do these results relate to our lives? [For example, the color of clothing we choose to wear outdoors in different seasons will affect our comfort. White would be more comfortable on a hot day and black on a cold day.]
7. Based on this experience, what new questions do you have? What different things do you want to try?
8. What are you wondering now?

Solar Mitts

Key Question

Which colors feel warmest in the sun?

Learning Goal

Students will:

use their sense of touch to compare how different colors absorb the sun's heat energy.

 © 2010 AIMS Education Foundation

Solar Mitts

Which colors feel warmest in the sun?

List of colors we will test:

Making the mitts
For each color, fold the construction paper in half to make a 6-inch by 9-inch rectangle. Tape or staple one short side and the long side across from the fold.

Using the mitts
- Put a colored mitt on each hand. Hold your hands in the sun.

- After about a minute, which mitt feels warmer?

- Have others in your group try the same colors. Do you agree?

- Keep comparing different colors until your group is ready to place the mitts in order from coolest to warmest.

- Draw the mitts on the line below and color them.

Coolest _____ **Warmest**

Solar Mitts

Connecting Learning

1. What was the source of heat for this activity?

2. Did the mitts of all colors feel the same? Explain.

3. Which colors were easy to place?

4. Which colors were harder to put into order? How did your group solve any disagreements?

5. How do our class results compare?

6. How do these results relate to our lives?

7. Based on this experience, what new questions do you have? What different things do you want to try?

8. What are you wondering now?

 © 2010 AIMS Education Foundation

Hot Pockets

Topic
Heat energy

Key Question
How does color affect temperature?

Learning Goal
Students will explore the effects of color on the absorption of heat energy from the sun.

Guiding Documents
Project 2061 Benchmarks
- *When warmer things are put with cooler ones, the warm ones lose heat and the cool ones gain it until they are all at the same temperature. A warmer object can warm a cooler one by contact or at a distance.*
- *Tables and graphs can show how values of one quantity are related to values of another.*
- *Keep records of their investigations and observations and not change the records later.*

NRC Standards
- *The sun provides the light and heat necessary to maintain the temperature of the earth.*
- *Heat can be produced in many ways, such as burning, rubbing, or mixing one substance with another. Heat can move from one object to another by conduction.*
- *Heat moves in predictable ways, flowing from warmer objects to cooler ones, until both reach the same temperature.*

NCTM Standards 2000*
- *Understand such attributes as length, area, weight, volume, and size of angle and select the appropriate type of unit for measuring each attribute*
- *Select and apply appropriate standard units and tools to measure length, area, volume, weight, time, temperature, and the size of angles*
- *Collect data using observations, surveys, and experiments*
- *Represent data using tables and graphs such as line plots, bar graphs, and line graphs*

Math
Measurement
   temperature
   time
Graphing

Science
Physical science
   heat energy
     radiation
    temperature

Integrated Processes
Observing
Controlling variables
Collecting and recording data
Comparing and contrasting
Interpreting data
Relating

Materials
For each group:
   2 thermometers with matching readings
   piece of cardboard, about 8 inches by 12 inches
   4-inch by 10-inch construction paper, 1 black and
     1 white

For the class:
   watch, clock, or timer
   tape

Background Information
The sun is the major source of heat energy on Earth. When the sun's rays strike the Earth by radiation, each surface reflects or absorbs a varying amount of heat energy. The more the sun's rays are absorbed, the more the temperature will rise.

Color is one of the factors affecting how many of the sun's rays are reflected or absorbed. Light colors reflect more of the sun's rays while dark colors absorb more. In fact, a good absorber appears black because it reflects very little light.

Whether the surface is dull or shiny, rough or smooth, or thin or thick also influences how much heat energy is absorbed. Matter that absorbs heat energy also emits or radiates heat energy. A grassy surface is suggested for this investigation because it emits less heat energy than, say, concrete. In addition, the cardboard acts as a partial insulator against heat transfer from the surface on which it is placed.

Black and white are used in this activity because they present the greatest contrast between dark and light. To isolate the variable of color, care must be taken to choose paper of the same weight and smoothness. The pockets should be the same size and placed on the same surface.

Management

1. To collect valid data, each pair of thermometers must have matching readings. However, pairs of thermometers may differ from each other. You may want to pre-match thermometers or have students find their own matching pair. Thermometers (item number 1976) are available from AIMS.

2. Gather materials in one place so groups can get what they need at the appropriate time. Make sure the black and white paper are of the same weight and smoothness as construction paper can vary, affecting the results.

3. Groups of three or four are suggested, depending on the availability of thermometers.

4. Plan to do this activity on a sunny day when the air is relatively calm. Winds may temper the results because of the wind chill factor. If possible, set the cardboard on a grassy surface not too close to buildings.

5. Instruct the students to read the temperature by positioning the eyes directly over the thermometer (90° angle) and pulling it out of the pocket just far enough to see. The thermometer bulb should not be exposed to the sun, as this may increase the reading. They should touch the thermometer only along its edges so their body heat does not affect the temperature.

6. Students may be tempted to alter data if unexpected results occur. Remind them that scientists record the data accurately and then look for reasons that might explain unexpected results. For example, temperatures that have been rising might fall during one reading if the sun temporarily goes behind a cloud or a breeze becomes stronger.

7. *Alternate approach*: Follow *Procedure* but consider having students design and construct their own tables and graphs instead of using the ones on the activity pages. This requires more thought as to what needs to be recorded and how best to represent it on a graph. A broken-line graph is most appropriate for data dealing with change over time; however, if your students are not ready for that format, use a bar graph.

Procedure

1. Ask, "What clothes colors do you wear in the summer?" (may yield a wide variety of responses but, in general, lighter colors dominate) "I wonder if there is any reason for this?"

2. Explain that they are going to investigate whether color has anything to do with how comfortable they feel, temperature-wise. They will be using black and white since they are the darkest and lightest colors. Give students the first activity page.

3. Instruct students on how to assemble their pockets, then have each group collect the materials from a designated location and prepare the pockets.

4. Ask, "How are we making this a fair test?" [the two papers for the pockets are the same weight and smoothness, the pockets are the same size and constructed in the same way, the thermometers' readings match at the beginning, both colors will be in the sun and on the same kind of surface, neither thermometer bulb will be directly exposed to the sun, etc.]

5. Direct students to predict how the temperatures inside the two pockets will compare. (There are three possibilities: The white pocket will be warmer than the black pocket, the two pockets will have the same temperatures, or the black pocket will be warmer than the white pocket.)

6. Take the class outside to a shady spot and have them lay the thermometers on the cardboard next to the pockets. Explain that the first temperature they will record is the air temperature. Meteorologists take this in the shade.

7. While waiting for the thermometers to stabilize, instruct students to observe and record the sky and general weather conditions using words such as sunny, clear, calm, light breeze, a few clouds, etc.

8. Once the red liquid in the thermometers stops moving, have students record the time and temperature for each thermometer. (The starting temperature should be the same for both.)

9. Direct groups to insert the thermometers in the pockets and place the cardboard in the sun.

10. Have students wait five minutes, marked by a watch, clock, or timer, and read the temperature for each thermometer by pulling it only partway out of the pocket (see *Management 5*).

11. Repeat every five minutes for a total of half an hour.

12. Return to the classroom and give students the graph page. Have them number the temperature increments (1° if possible), label the times, make a key, graph the data, and give the graph a title.

13. Discuss what the graph shows, its story. Have students write about the results and discuss how it relates to them.

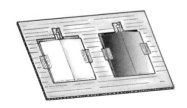

 © 2010 AIMS Education Foundation

Connecting Learning

1. What story does the graph tell? [Temperatures rise the fastest in the first five minutes. Black causes the temperature to rise more than white. At the end of half an hour, the black temperature was ____ degrees higher than the white temperature.]

2. Compare the black and white temperature differences from the second reading on. What kind of pattern do you notice? [The spread between the two temperatures gradually becomes bigger. (The difference may quit widening at some point.)]

3. In what ways might color affect your comfort level, temperature-wise? [dark *houses* absorb more heat energy and make it more difficult to cool in hot, sunny weather; dark-colored *cars* get warmer than light-colored cars; white or very pale-colored *clothing* is cooler to wear, while dark-colored clothing will keep you warmer in direct sun, etc.]

4. How would color affect how you dress in the summer? ...in the winter?

5. What are you wondering now?

Extensions

Absorbing radiated heat

1. Artists, fashion designers, interior decorators, and others talk about warm colors and cool colors. What are some of the cool colors? [blue, green, purple] What are some of the warm colors? [yellow, pink, orange, red] Do the words *cool* and *warm* also refer to their effect on temperature? In other words, do cool colors also cause you to stay cooler in the sun? Do warm colors make you warmer?

   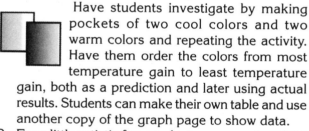 Have students investigate by making pockets of two cool colors and two warm colors and repeating the activity. Have them order the colors from most temperature gain to least temperature gain, both as a prediction and later using actual results. Students can make their own table and use another copy of the graph page to show data.

2. For a little artistic fun, students may want to experiment with multi-colored pockets using stripes, polka-dots, etc.

3. For those in areas where snow may occur, ask students where they see snow melting first. [Examples: A thin layer of snow on asphalt, a dark color, tends to melt before a thin layer of snow on concrete, a light color. Dirty snow melts faster than clean snow.]
   On a calm, sunny day, place pieces of black and white construction paper directly on the snow. Wait at least half an hour. Remove the papers and observe or measure the depth of the indentation in the snow. Under which paper did the snow melt the most? [Black.] Why? [Black absorbs more radiated heat which, in turn, melts the snow more rapidly.]

4. *Hot Pockets* compares the temperature of the air inside paper pockets. A related investigation can be done with water. Tightly wrap two empty soup cans, one with black construction paper and the other with white. Measure and fill each with the same amount of water. Take a beginning reading in the shade and then put in the sun. It is best to do this in the first part of the morning or the later part of the afternoon, when the sun's rays will strike the sides of the cans. Take readings 15 minutes apart for about two hours. (Water absorbs as well as releases heat energy slowly.) Compare results.

Emitting radiated heat

5. After leaving the black and white pockets in the sun for half an hour, continue the investigation by moving the pockets into the shade and again recording temperatures, this time at one-minute intervals. Now students will be comparing how quickly black and white give off or emit heat energy.

6. Run the paper doll on card stock. Have students use construction paper for making seasonal clothes. Place the paper doll with clothes in direct sunlight. Insert a thermometer between the clothes and card stock cutout to see how the clothes affect the absorption of solar energy.

*   Reprinted with permission from *Principles and Standards for School Mathematics,* 2000 by the National Council of Teachers of Mathematics. All rights reserved.

Hot Pockets

Key Question

How does color affect temperature?

Learning Goal

Students will:

explore the effects of color on the absorption of heat energy.

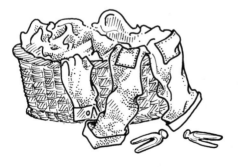

 © 2010 AIMS Education Foundation

Hot Pockets

How does color affect temperature?

Fold each paper in half lengthwise, then open. Fold in half across the width and tape both sides to a piece of cardboard. The lengthwise crease should make a slight hill so the thermometer can slide in easily.

Place in the sun and record temperatures every _____ minutes.

How do you think the temperatures will compare?

Time	Temperature (°C)	
	Black	White

Sky and weather:

          © 2010 AIMS Education Foundation

Hot Pockets

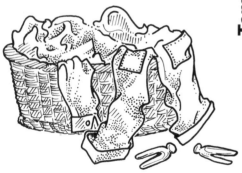

What is the story of the graph?

Temperature (°C)

Time

Paper Doll Patterns

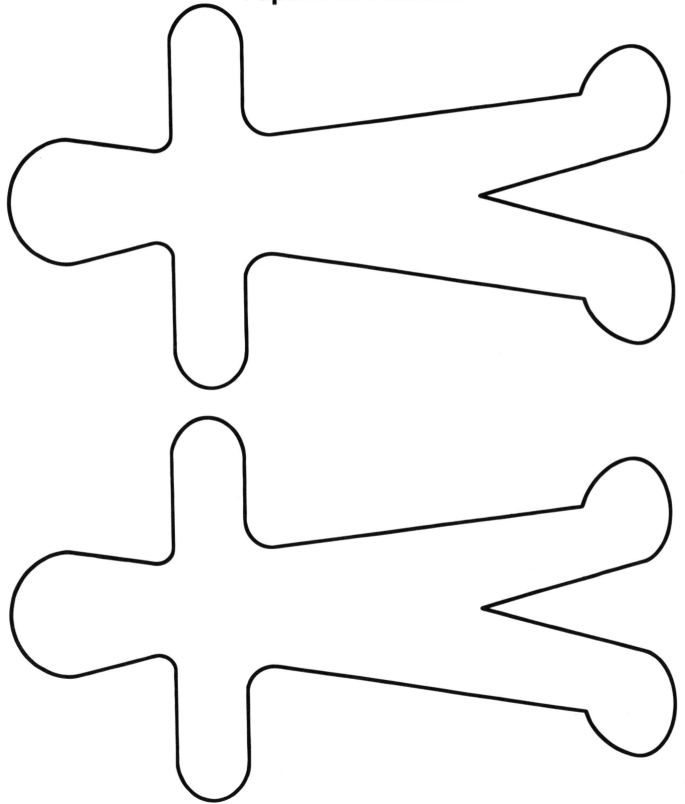

Connecting Learning

1. What story does the graph tell?

2. Compare the black and white temperature differences from the second reading on. What kind of pattern do you notice?

3. In what ways might color affect your comfort level, temperature-wise?

4. How would color affect how you dress in the summer? …in the winter?

5. What are you wondering now?

HEAT and COLOR

Topic
Heat energy

Key Question
What colors—dark or light—absorb heat better?

Learning Goals
Students will:
* compare the temperatures of water in black and white cups that have been exposed to direct sunlight, and
* draw conclusions about which color absorbs heat better.

Guiding Documents
Project 2061 Benchmark
* *Some materials conduct heat much better than others. Poor conductors can reduce heat loss.*

NRC Standards
* *Objects have many observable properties, including size, weight, shape, color, temperature, and the ability to react with other substances. Those properties can be measured using tools, such as rulers, balances, and thermometers.*
* *Heat can be produced in many ways, such as burning, rubbing, or mixing one substance with another. Heat can move from one object to another by conduction.*

NCTM Standards 2000*
* *Select and apply appropriate standard units and tools to measure length, area, volume, weight, time, temperature, and the size of angles*
* *Represent data using concrete objects, pictures, and graphs*
* *Describe parts of the data and the set of data as a whole to determine what the data show*

Math
Measurement
     temperature
Graphing
     bar graph

Science
Physical science
     heat energy
          temperature

Integrated Processes
Observing
Communicating
Collecting and recording data
Interpreting data
Drawing conclusions

Materials
Plastic cups, 10 oz
Black construction paper
White construction paper
Two thermometers (see *Management 2*)
Water

Background Information
    Black or dark colors absorb more solar energy than light colors, which reflect much of the solar energy. Light that is absorbed is changed to heat energy. Temperature is a measure of heat energy. The black cups will collect more of this energy so the water temperature inside will rise faster than the water temperature inside the white cups.

Management
1. Make one cup black by covering its exterior with black construction paper. Cover the outside of the other cup with white construction paper.
2. Before copying the student page, write numbers on the thermometers that are appropriate for either Celsius or Fahrenheit. Thermometers (item number 1976) are available from AIMS.
3. Let the water come to room temperature before beginning the activity.
4. Do this activity on a warm, sunny day.

Procedure
1. Place two pieces of construction paper, one black and one white, in the sun. After a short period, let the students feel each piece of paper. Ask them if there is a difference in the temperatures of the papers.
2. Discuss why it is beneficial to wear light colored clothes in the summer and darker colors in the winter.
3. Pour equal amounts of water into the black and white cups.
4. Put a thermometer in each cup. Have students read the thermometers and record the beginning temperature on their activity pages. Remove the thermometers.

5. Put the cups outside in the sunshine for 30 minutes.
6. Take the temperature of the water in each cup again. Have students record the temperatures on their student pages.
7. Do the same procedure after another 30 minutes and discuss the results.

Connecting Learning

1. What was the difference in the temperatures of the two cups after 30 minutes? ...60 minutes?
2. Why do you think there was a difference?
3. What do you think would happen if the cups had been placed outside in the shade?
4. What conclusion can you draw about dark colors and light colors that are in the sun?
5. What do you think would happen if we tried another color on the outside of the cup? What if we put aluminum foil on the outside of the cup?
6. What do you think would happen if you covered the tops of the cups with plastic wrap?
7. What are you wondering now?

* Reprinted with permission from *Principles and Standards for School Mathematics, 2000* by the National Council of Teachers of Mathematics. All rights reserved.

HEAT and COLOR

Key Question

What colors—dark or light—absorb heat better?

Learning Goals

Students will:

- compare the temperatures of water in black and white cups that have been exposed to direct sunlight, and

- draw conclusions about which color absorbs heat better.

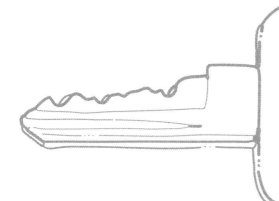

HEAT and COLOR

Put two cups filled with water outside in the sun. Measure the temperature of the water. Color in the thermometers below.

Black cup

White cup

Start | After 30 minutes | After 60 minutes

Start | After 30 minutes | After 60 minutes

What did you find out?

HEAT and COLOR

Connecting Learning

1. What was the difference in the temperatures of the two cups after 30 minutes? ...60 minutes?

2. Why do you think there was a difference?

3. What do you think would happen if the cups had been placed outside in the shade?

4. What conclusion can you draw about dark colors and light colors that are in the sun?

Connecting Learning

5. What do you think would happen if we tried another color on the outside of the cup? What if we put aluminum foil on the outside of the cup?

6. What do you think would happen if you covered the tops of the cups with plastic wrap?

7. What are you wondering now?

Curly Cue

Topic
Heat energy

Key Question
How does heat affect the air?

Learning Goal
Students will discover that heat energy causes the air to move upward.

Guiding Documents
Project 2061 Benchmark
- *Things that give off light often also give off heat. Heat is produced by mechanical and electrical machines, and any time one thing rubs against something else.*

NRC Standards
- *Heat can be produced in many ways, such as burning, rubbing, or mixing one substance with another. Heat can move from one object to another by conduction.*
- *Energy is a property of many substances and is associated with heat, light, electricity, mechanical motion, sound, nuclei, and the nature of a chemical. Energy is transferred in many ways.*

Science
Physical science
  heat energy
Safety

Integrated Processes
Observing
Comparing and contrasting
Interpreting data

Materials
Heat sources (see *Management 2*)
Thread
Scissors
Transparent tape
Student pages
Small paper plates, optional (see *Management 3*)

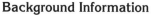

Background Information
Heat energy moves. It spreads from one place to another in several ways. One way it moves is in air currents. When air is heated, as it is in this activity, it becomes less dense and moves upward. The effects of the upward-moving warm air can be observed on the paper coil. The more heat that is generated, the faster the air moves and the faster the coil will spin. Currents are evidence of convection at work. (When a heated substance itself moves (air in this case), heat energy is being transferred by convection.) Both gases and liquids can have currents. These currents are evidence of heat energy's ability to create motion and cause change.

Management
1. Safety is always a consideration when heat is involved. Choose heat sources where there is no flame. Supervise students while they are using these objects.
2. Set up three areas in the room, one with a heat source turned off, one with a source generating low heat (such as a low-watt incandescent or fluorescent light bulb), and one where medium to high heat is generated (100-watt incandescent bulb, toaster, hot plate, etc.). Students can rotate through the three areas.
3. Students can cut the coil from the student page or cut their own from thin paper plates.

Procedure
1. Distribute the first student page, scissors, thread, and tape, and have students assemble the paper coil. Provide assistance as needed.
2. Explain that students will be testing the ability of heat energy to cause motion by holding the paper coil over three different heat sources. Ask them what safety rules should be established with regards to the heat sources.
3. Demonstrate how to hold the coil above a heat source so that it can spin freely.
4. Have students take turns holding first their hands and then their coils in various positions around and above a heat source that has not been turned on.
5. Distribute the second student page and instruct them to record the heat source, the amount of heat [none], and their observations [nothing happened, it felt like the rest of the air in the room].

 © 2010 AIMS Education Foundation

6. Direct students to repeat this process for the other two heat sources. The amount of heat for the remaining two sources should be recorded as low and high, respectively.
7. Invite students to observe the relative speed of the moving coil and note the position in which the coil detected the warm air. [above the heat source]
8. Lead a class discussion about how heat energy affects the air and instruct students to write what they have learned.

Connecting Learning
1. What do you think caused the coil to move? [warm air rising]
2. What caused the air to be warm? [heat energy from the light, toaster, etc.]
3. Where in relation to the heat source did you notice air moving? [above it] How did the air feel? [warm]
4. How did your observations of the three heat sources compare? [The air did not move where there was no heat. The air moved upward over the heat sources that were on. The lower the heat, the more slowly the air moved.]
5. Based on the results of this investigation, do you think the temperature would be warmer near the ceiling or near the floor? Justify your response.
6. Where in your house would you want to sleep during the summer? (Mention possibilities such as upstairs or downstairs, top bunk or lower bunk.)
7. What other changes have you observed that are caused by heat energy?
8. What are you wondering now?

Extension
Find a third heat source that is either hotter than the hottest source used or in between the two sources used. Have students predict how the coils will move and test their predictions.

Curly Cue

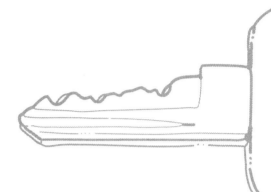

Key Question

How does heat affect the air?

Learning Goal

Students will:

discover that heat energy causes the air to move upward.

 © 2010 AIMS Education Foundation

Curly Cue

Cut out the spiral along the dark lines. Tape a piece of thread to the dot in the center. Hold the spiral so it can move freely.

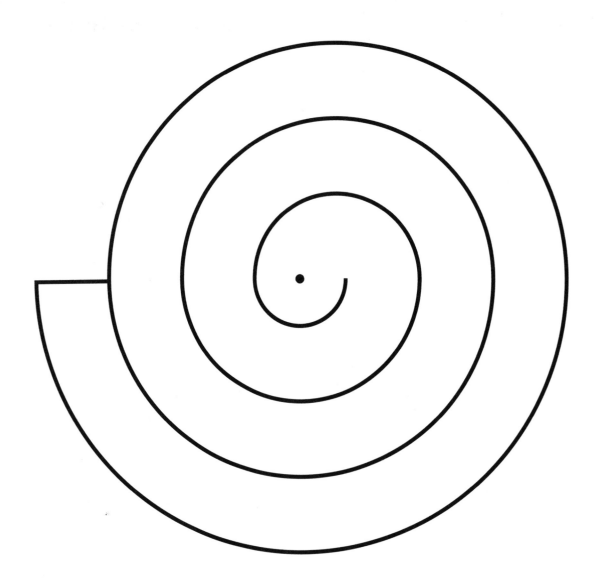

© 2010 AIMS Education Foundation

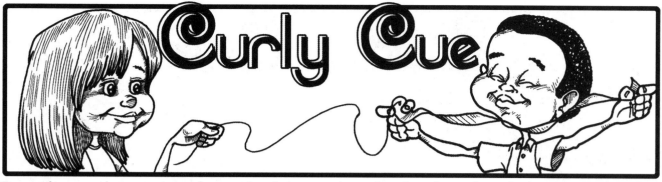

Record your observations for each of the stations. Look at how fast the coil moves and where it responded to the warm air.

Heat Source: _____ **Amount of Heat:** _____

Observations:

Heat Source: _____ **Amount of Heat:** _____

Observations:

Heat Source: _____ **Amount of Heat:** _____

Observations:

What did you learn from this investigation?

Curly Cue

Connecting Learning

1. What do you think caused the coil to move?

2. What caused the air to be warm?

3. Where in relation to the heat source did you notice air moving? How did the air feel?

4. How did your observations of the three heat sources compare?

5. Based on the results of this investigation, do you think the temperature would be warmer near the ceiling or near the floor? Justify your response.

Curly Cue

Connecting Learning

6. Where in your house would you want to sleep during the summer?

7. What other changes have you observed that are caused by heat energy?

8. What are you wondering now?

Heat From FRICTION

Topic
Friction

Key Question
What happens when you rub two objects together?

Learning Goal
Students will realize that rubbing two surfaces together produces heat.

Guiding Documents
Project 2061 Benchmark
• *Things that give off light often also give off heat. Heat is produced by mechanical and electrical machines, and any time one thing rubs against something else.*

NRC Standard
• *Heat can be produced in many ways, such as burning, rubbing, or mixing one substance with another. Heat can move from one object to another by conduction.*

Science
Physical science
  heat energy
    friction

Integrated Processes
Observing
Communicating
Comparing and contrasting

Background Information
Friction is the force that opposes motion. Friction always changes work to heat. Two objects, when rubbed together, cause friction, a speeding up of the molecules in the objects. If you rub your palms together, you can feel heat caused by friction.

Materials with rough surfaces can create more friction and resulting heat than those with smooth surfaces. Friction can be produced on smooth surfaces by rubbing them rapidly for a long time, but it will take longer than with rough surfaces and the resulting heat will be less.

Procedure
1. Write the word *friction* on the board and discuss that friction occurs when two objects are rubbed together. Tell students that friction produces heat.
2. Direct students to put their hands on their faces. Have them discuss how their hands feel.
3. Tell them to press the palms of their hands together and rub them fast and hard while they count to 30. Have them put their hands on their faces again. Ask them how their hands feel now. Why? [They are warmer because the friction created heat.]
4. Let the students' hands cool off and then repeat. Ask them if they got the same results.
5. Have the students rub their hands over their desktops, on top of their jeans, over the surface of a book, etc. Each time, have them place their hands on their faces and describe what they feel.

Connecting Learning
1. What is friction?
2. Name some ways that you can create friction.
3. Does it make a difference in the amount of heat produced if you rub your hands fast or slow? Explain.
4. Do we need friction? Explain. [Yes. It is friction that makes the brakes stop our bikes and cars. It is friction that keeps our shoes from slipping on the floor.]
5. How is friction used to stop motion?
6. What are you wondering now?

Extensions
1. Rubbing two inflated balloons together will create friction and heat. One balloon will eventually pop. Ask the students why rubbing the balloons together made one pop. [Because the friction between the two balloons made one balloon so hot that part of it melted.]
2. Try rubbing two sheets of fine sandpaper together. There will be enough heat produced that you will not want to put your hands on them.
3. Friction is often unwanted heat energy. Explain that machines with moving parts encounter friction when parts touch. Oil is added to reduce friction. Have the students repeat the *Heat From Friction* activity with lotion on their hands. How do the results differ?

Heat From FRICTION

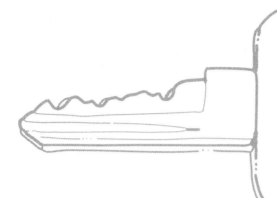

Key Question

What happens when you rub two objects together?

Learning Goal

Students will:

realize that rubbing two surfaces together produces heat.

Heat From FRICTION

Connecting Learning

1. What is friction?

2. Name some ways that you can create friction.

3. Does it make a difference in the amount of heat produced if you rub your hands fast or slow? Explain.

4. Do we need friction? Explain.

5. How is friction used to stop motion?

6. What are you wondering now?

HOT STUFF

Topic
Heat energy

Key Question
How can heat be produced?

Learning Goals
Students will:
- explore a variety of ways heat is produced, and
- identify sources of heat.

Guiding Documents
Project 2061 Benchmark
- *Things that give off light often also give off heat. Heat is produced by mechanical and electrical machines, and any time one thing rubs against something else.*

NRC Standard
- *Heat can be produced in many ways, such as burning, rubbing, or mixing one substance with another. Heat can move from one object to another by conduction.*

Science
Physical science
   heat energy

Integrated Processes
Observing
Classifying
Collecting and recording data
Comparing and contrasting
Interpreting data
Drawing conclusions

Materials
Exploration 1: hand lotion
Exploration 2: hair dryer, electrical outlet
Exploration 3: desk lamp, electrical outlet
Exploration 4: candle, match (see *Management 3*)
Exploration 5: block of wood, pieces of sandpaper
Exploration 6: wire hanger
Rubber bands, #19
Safety glasses

Background Information
   One type of energy students can easily explore is heat energy. It is produced in many ways. They can begin to look at objects and actions that give off heat like lights, hair dryers, the sun, rubbing, bending things, etc. This lesson is intended to give students a general idea of ways heat is produced from rubbing, bending, burning, and electricity.

Management
1. The procedures and directions for each station are listed on the station cards. Copy the cards for each station on card stock and laminate for extended use.
2. Use a candle in a jar. Make sure there are no flammable materials nearby. Only an adult supervisor should light the candle.

Procedure
1. Ask the *Key Question* and state the *Learning Goals*.
2. Have students fold and assemble their rubber band books. Allow time for reading and discussing.
3. Explain to the students that they will be exploring heat at several stations. Inform the students that they will be bending a wire. Ask for their help in determining what safety precautions and equipment they will need to protect their eyes. (Safety glasses should be worn.) Discuss what safety precautions are also necessary at each of the exploration stations.
4. Allow the students time to rotate through each station. Have them record their observations on the student sheet.
5. End with a period of discussion during which results from each station are covered.

Connecting Learning
1. What are some ways heat is produced?
2. Why is rubbing your hands together useful or helpful? How is it similar to running in place?
3. What other things get hot when you rub them together? [sandpaper and wood, brakes on a bicycle tire, etc.]
4. Which of the stations did you find most surprising? Why?
5. What are some of the benefits and dangers of heat being produced?
6. What would life be like if we didn't have these kinds of heat-producing situations?
7. In what ways has this lesson changed the way you think about heat? Explain.
8. What are you wondering now?

HOT STUFF

Key Question

How can heat
be produced?

Learning Goals

Students will:

- explore a variety of ways heat is produced, and
- identify sources of heat.

© 2010 AIMS Education Foundation

1

THE HEAT IS ON

8

BYE!

Heat energy is all around us. It powers our vehicles, warms our homes, and cooks our food.

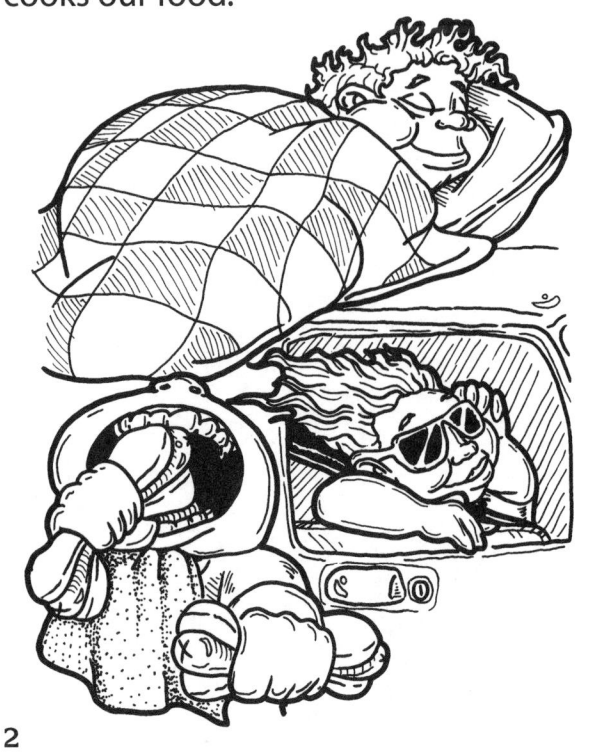

2

Heat also comes from burning. On a cold winter day, we start a fire in the fireplace to warm our houses.

7

5

Heat can come from electricity. We use electricity to cook our food, warm our schools, and light our homes when there is no sunlight.

Heat can come from the sun. The heat that comes from the sun is called solar energy. The sun warms the Earth and gives us daylight.

Heat can come from objects rubbing against each other, or friction. We rub our hands together when they are cold. The friction creates heat.

There are many sources of heat.

6

3

ENERGY EXPLORATIONS

© 2010 AIMS Education Foundation

HOT STUFF

Follow the directions on each station card. Record your observations. Indicate the energy source in each case.

Station	Observation 1	Observation 2	Energy Source
1. Hands			
2. Hair dryer			
3. Lamp			
4. Candle			
5. Wood and sandpaper			
6. Wire hanger			

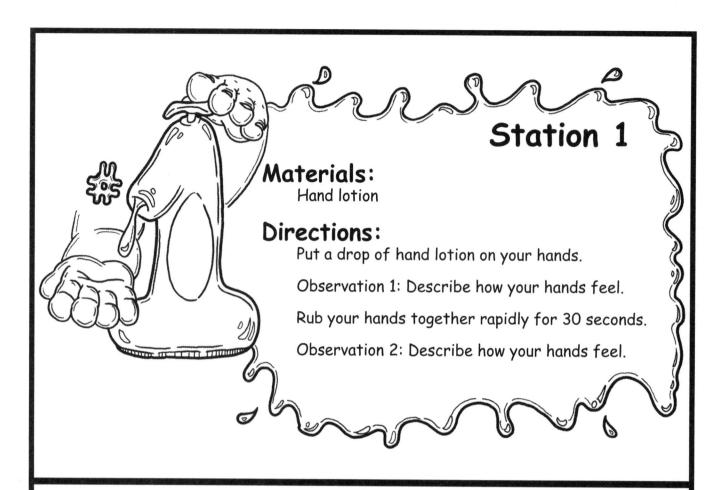

Station 1

Materials:
Hand lotion

Directions:
Put a drop of hand lotion on your hands.

Observation 1: Describe how your hands feel.

Rub your hands together rapidly for 30 seconds.

Observation 2: Describe how your hands feel.

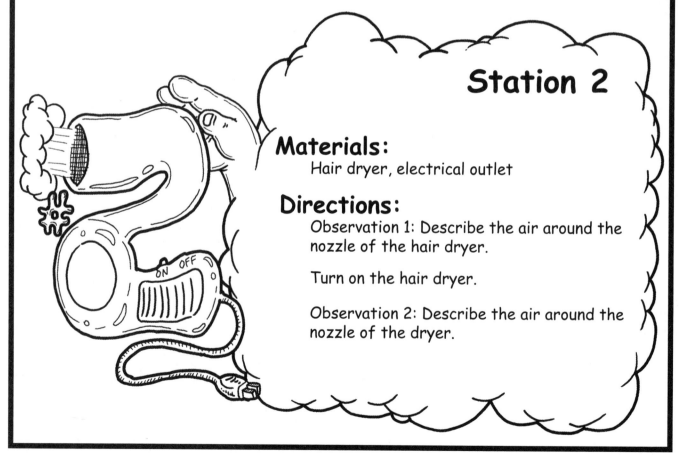

Station 2

Materials:
Hair dryer, electrical outlet

Directions:
Observation 1: Describe the air around the nozzle of the hair dryer.

Turn on the hair dryer.

Observation 2: Describe the air around the nozzle of the dryer.

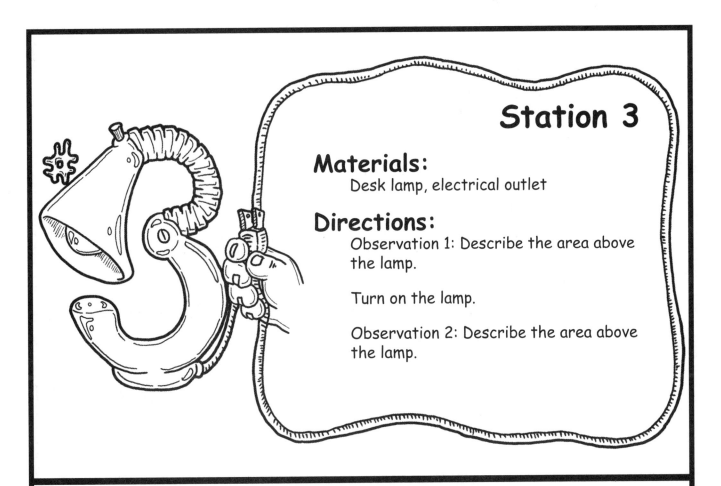

Station 3

Materials:
Desk lamp, electrical outlet

Directions:
Observation 1: Describe the area above the lamp.

Turn on the lamp.

Observation 2: Describe the area above the lamp.

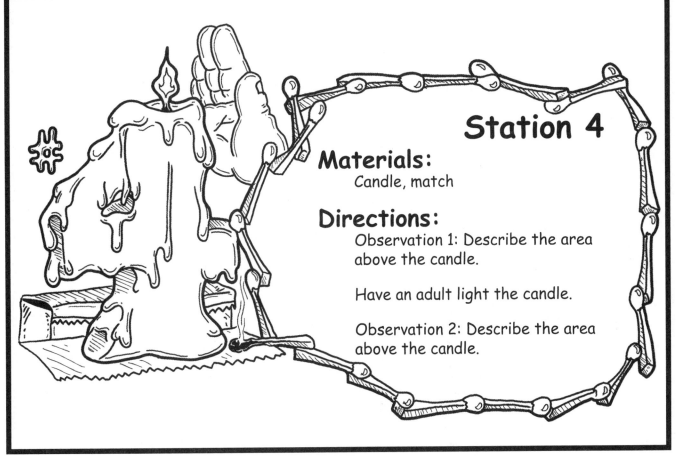

Station 4

Materials:
Candle, match

Directions:
Observation 1: Describe the area above the candle.

Have an adult light the candle.

Observation 2: Describe the area above the candle.

Station 5

Materials:
Block of wood, sandpaper

Directions:
Observation 1: Describe your sandpaper and the block of wood.

Rub your sandpaper on the wooden block rapidly for 30 seconds.

Observation 2: Describe your sandpaper and the block of wood.

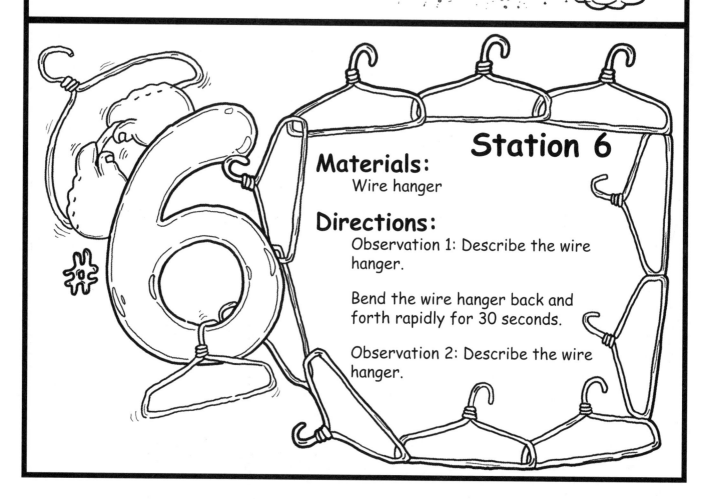

Station 6

Materials:
Wire hanger

Directions:
Observation 1: Describe the wire hanger.

Bend the wire hanger back and forth rapidly for 30 seconds.

Observation 2: Describe the wire hanger.

HOT STUFF

Connecting Learning

1. What are some ways heat is produced?

2. Why is rubbing your hands together useful or helpful? How is it similar to running in place?

3. What other things get hot when you rub them together?

4. Which of the stations did you find most surprising? Why?

5. What are some of the benefits and dangers of heat being produced?

Connecting Learning

6. What would life be like if we didn't have these kinds of heat-producing situations?

7. In what ways has this lesson changed the way you think about heat? Explain.

8. What are you wondering now?

 © 2010 AIMS Education Foundation

No matter where we put the thermometer, in ice water or boiling soup, the energy will flow in or out and make the thermometer the same temperature as the substance it touches.

Heat and temperature are related but are not the same thing. Most people are familiar with the idea of temperature. The temperature of our bodies is important to health. Recipes tell us the temperature of the air in the oven for baking. The weather forecaster gives us the temperature of today's weather. Our experience tells us that the warmer something feels, the higher the temperature is likely to be.

There are different ways to put numbers on a thermometer. The scale that most scientists use is called Celsius. The one we use everyday is Fahrenheit. Both measure temperature, they just use different scales—sort of like a yardstick and a meter stick that are used to measure length.

8

Heat Energy and Temperature

1

© 2010 AIMS Education Foundation

If we put the end of a thermometer in our mouths, heat energy will flow from our mouths to the thermometer and it will come to the same temperature as our bodies.

Heat energy flows from warmer to cooler. If two objects or materials are put into contact, and we wait until all changes stop, the objects will be at the same temperature.

If a thermometer is left in a room, it will come to the same temperature as the air.

A thermometer is an instrument whose size, shape, or some feature changes when its temperature changes so that it can be used to measure temperature. The most common thermometers are those that have an expanding column of mercury or colored alcohol. A thermostat has a strip of two types of metal that curls and uncurls to measure temperature.

© 2010 AIMS Education Foundation

Make a Thermometer

Topic
Thermometers

Key Question
How does a thermometer work?

Learning Goal
Students will observe colored liquid rise in a tube and relate it to the working of a thermometer.

Guiding Documents
Project 2061 Benchmark
- *Human beings have made tools and machines to sense and do things that they could not otherwise sense or do at all, or as quickly, or as well.*

NRC Standard
- *Objects have many observable properties, including size, weight, shape, color, temperature, and the ability to react with other substances. Those properties can be measured using tools, such as rulers, balances, and thermometers.*

Science
Physical science
 heat energy
  temperature

Integrated Processes
Observing
Comparing and contrasting
Relating

Materials
Glass bottle
Drinking straw
Clay
Red food coloring
Liter boxes or other transparent containers
Hot water
Cold water

Background Information
The fluid in a thermometer rises in its column because the fluid expands when it is heated. Because there is nowhere else for the fluid to go, it rises in the column. When the fluid cools, it falls.

In this activity, students will watch a colored fluid rise in a drinking straw as the water is warmed. If the seal of clay is tight, the only place for the water to go is up the straw. This experience is then related to the rising fluid in a thermometer.

Management
1. This activity can be done as a demonstration or set up in a center for students. Hot tap water will get good results. You may want to have a pitcher of ice water so the temperature differences of the water is greater.
2. Add red food coloring to the cold water to replicate the color of the liquid in most thermometers students will use.
3. Liter boxes (item number 1913) are available from AIMS.

Procedure
1. Ask the *Key Question* and state the *Learning Goal*.
2. Distribute the student page. Go over the directions with the students.
3. Caution them that they will need to make sure the clay seal has no air holes.
4. Set up the activity. Have students record their observations.

Connecting Learning
1. What happened to the liquid in the straw?
2. Did anyone have difficulty getting the liquid to go up the straw? What was the reason it didn't? How did you fix it?
3. How is this like a thermometer? How is it different?
4. What happens to the liquid in a thermometer when it is cold?
5. What happens to the liquid when it is hot?
6. What are you wondering now?

Make a Thermometer

Key Question

How does a thermometer work?

Learning Goal

Students will:

observe colored liquid rise in a tube and relate it to the working of a thermometer.

 © 2010 AIMS Education Foundation

Make a Thermometer

You will need:

- straw
- glass bottle
- clay
- food coloring
- 2 deep pans
- hot water
- cold water

Do this:

- Pour cold water into the bottle. Add a few drops of food coloring.
- Put a straw about halfway down into the water.
- Mold the clay around the top of the bottle. Make a tight seal.
- Put the bottle into a deep pan.
- Pour hot tap water into the pan.
- Watch the water in the straw.

1. Describe what you observe.

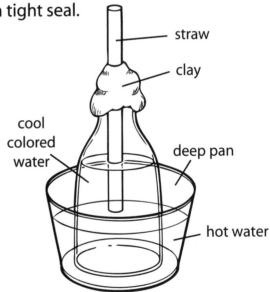

straw

clay

cool colored water

deep pan

hot water

- Put the bottle into the other deep pan.
- Pour cold water into the pan.

2. Describe what you observe.

Make a Thermometer

Connecting Learning

1. What happened to the liquid in the straw?

2. Did anyone have difficulty getting the liquid to go up the straw? What was the reason it didn't? How did you fix it?

3. How is this like a thermometer?

4. What happens to the liquid in a thermometer when it is cold?

5. What happens to the liquid when it is hot?

6. What are you wondering now?

 © 2010 AIMS Education Foundation

When you step on the carpet, the heat from your feet does not move as easily through the carpet. It feels warmer than the tile, but it isn't. The heat from your feet isn't moving as fast.

means carry.
The word conduct
called conduction.
is
Did you know that heat energy moves? When objects are touching each other, heat moves from one to the other. It moves from hotter objects to cooler ones. This movement of heat energy is called conduction. The word conduct means carry.

You can use your sense of touch to help you decide if something is a good conductor or a good insulator.

Heat Energy: Conduction

© 2010 AIMS Education Foundation

On a cold winter morning, the carpet in your house feels warmer than the tile on your bare feet. The carpet and the tile are the same temperature. Why do they feel different?

Materials that don't conduct heat energy well are called *insulators*.

The tile is a conductor. The carpet is an insulator. Your body temperature is warmer than the tile. When you step on the tile, the heat from your feet moves into the tile. The heat energy flows easily.

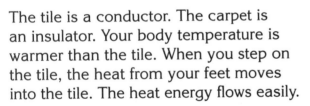

Some materials conduct heat energy more easily than others. These materials are called *conductors*.

VERY HOT!

Snake Warmers

Topic
Heat conduction

Key Question
How can you quantify the transfer of heat from a hot object to a cold object?

Learning Goals
Students will:
- make "snakes" from knee-high stockings,
- allow these snakes to bask on hot water bottle "rocks," and
- measure the temperatures of the snakes over time.

Guiding Documents
Project 2061 Benchmarks
- *When warmer things are put with cooler ones, the warm ones lose heat and the cool ones gain it until they are all at the same temperature. A warmer object can warm a cooler one by contact or at a distance.*
- *Tables and graphs can show how values of one quantity are related to values of another.*

NRC Standards
- *Heat moves in predictable ways, flowing from warmer objects to cooler ones, until both reach the same temperature.*
- *Objects have many observable properties, including size, weight, shape, color, temperature, and the ability to react with other substances. Those properties can be measured using tools, such as rulers, balances, and thermometers.*
- *Employ simple equipment and tools to gather data and extend the senses.*

NCTM Standards 2000*
- *Select and apply appropriate standard units and tools to measure length, area, volume, weight, time, temperature, and the size of angles*
- *Collect data using observations, surveys, and experiments*
- *Represent data using tables and graphs such as line plots, bar graphs, and line graphs*

Math
Measurement
    temperature
Graphing
    line graph

Science
Physical science
    heat energy
        conduction
Life science
    cold-blooded animals

Integrated Processes
Observing
Comparing and contrasting
Collecting and recording data
Relating

Materials
Knee-high stockings (see *Management 1*)
Paper towels
Permanent markers
Thermometers
1-gallon water bottles (see *Management 2*)
Hot tap water
Student pages

Background Information
One way in which heat energy travels is by conduction. This happens when objects are touching and the heat flows from the hotter object to the cooler one. We experience this constantly in our daily lives. In the winter, we may drink a cup of hot cocoa. While the liquid warms our insides as we drink it, our hands are warmed by holding the cup. The heat energy moves from the warmer cup to our colder hands. If we hold it long enough, the cup and our hands will eventually come to the same temperature, and the cup will no longer feel warm.

The principles of heat conduction are used by cold-blooded animals to increase their internal temperatures. Lizards and snakes can often be seen basking on rocks, especially in morning hours before the air has warmed significantly. The rocks are warmer than their bodies, so heat is conducted from the rocks to them. This is the situation that students will simulate in this activity as they make "snakes" from knee-high nylons and place them on hot water bottle "rocks" to bask.

Management
1. Buy a box of cheap knee-high nylon stockings. Select the lightest color available—nude or off white—so that students can easily read a thermometer through the nylon. Each group of students needs one knee-high.

2. You will need large one-gallon water bottles that have screw-on caps. The bottles need to be rectangular in shape so that they can be laid on their sides without rolling and have a flat surface on which students can place their snakes. If this kind of bottle is not available, two-liter bottles can be used, but a book must be placed on either side to prevent rolling. Another option is to use large rubber hot-water bottles. Each group of students needs one water bottle.

3. You will need a source of hot tap water. If you do not have hot water in your classroom, plan to go to the kitchen or another location where students will be able to fill their water bottles with hot water from a tap.

4. School-issue paper towels will work for stuffing the snakes. If you want to give students more colorful options, you can use colored tissue paper.

Procedure

1. Ask students to think of a time when they have had cold hands. Invite them to share some of the things they did to warm up their hands.

2. If no one mentions it, say that one way you like to warm your hands is by holding a mug of hot cocoa. Discuss why this (and any other examples that students gave relating to heat conduction) works. [Heat moves from hotter objects to cooler ones. The mug of cocoa is hotter than your hands, so some of that heat moves into your hands until your hands and the mug reach the same temperature.]

3. Challenge students to think of some real-world examples of animals using heat conduction to warm themselves. Discuss how snakes, lizards, and other cold-blooded animals will bask on rocks to warm themselves. Relate this situation to holding a mug of cocoa so that students recognize that the same principle is at work. (The rock is warmer than the snake's body, so heat is conducted from the rock to the snake.)

4. Explain that students are going to make some "snakes" out of knee-high stockings and put them on some hot "rocks" to see how much the snakes are warmed by the rocks.

5. Divide students into groups and distribute the materials for the snakes and the page of instructions. Allow time for groups to make and decorate their snakes.

6. Distribute the student page and have students find the starting temperatures of their snakes and record this data.

7. Give each group a water bottle to use for its rock. Have them fill these bottles with hot water all the way to the top and screw on the lids so that the bottles will not leak. Instruct them to use paper towels to dry the outsides of the bottles.

8. Have each group return to its area and place the snake on the water bottle. After two minutes, have them record the temperature of the snake. Continue to have them record the snake's temperature at two-minute intervals for 10 minutes.

9. Invite students to remove their snakes from their rocks and to feel the undersides of the snakes.

10. Instruct students to graph the data and discuss what it shows. (They will need to decide on a scale for the graph based on the data they collected.)

11. Close with a final time of class discussion and sharing.

Connecting Learning

1. What is heat conduction? [the flow of heat from a hotter object to a cooler object touching it]

2. What are some ways that you experience heat conduction in your day-to-day life? [warming hands by holding hot drinks, sleeping with an electric blanket in winter, getting in a warm bath or shower, etc.]

3. Does experiencing heat conduction always have to be you being warmed by something hotter than your body? [No.] Explain. [When you are hot, you can conduct heat away from your body by jumping in a swimming pool, putting a cool washcloth on your forehead, etc.]

4. What was the temperature of your group's snake before you put it on the rock?

5. What happened to the snake's temperature after two minutes? …four minutes? …etc.?

6. How did your data compare to the data collected by other groups? What might be some reasons for any differences? [thermometers that don't have matching readings, differences in location of thermometer relative to rock, measurement error, etc.]

7. How did your snake feel when you picked it up after it had been on the rock for 10 minutes? [warm]

8. What does this tell you about what was happening? [Heat energy from the rock was being conducted into the snake and the surrounding air.]

9. Do you think the temperature of the snake would have continued to go up if we had continued to measure for more than 10 minutes? Explain. [The temperature will increase to a point, then it will level off and eventually decrease as the rock (and the snake) comes to room temperature.]

10. What would happen if we left our snakes on the rocks overnight? [The snake and the rock will eventually come to room temperature as all of the heat energy from the hot water is conducted to the cooler surrounding air.]

11. What are you wondering now?

*   Reprinted with permission from *Principles and Standards for School Mathematics*, 2000 by the National Council of Teachers of Mathematics. All rights reserved.

                    © 2010 AIMS Education Foundation

Snake Warmers

Key Question

How can you quantify the transfer of heat from a hot object to a cold object?

Learning Goals

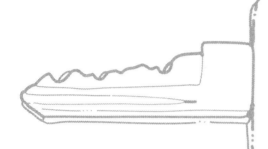

Students will:

- make "snakes" from knee-high stockings,
- allow these snakes to bask on hot water bottle "rocks," and
- measure the temperatures of the snakes over time.

Snake Warmers

You will be making a "snake" from a knee-high stocking.

You need:

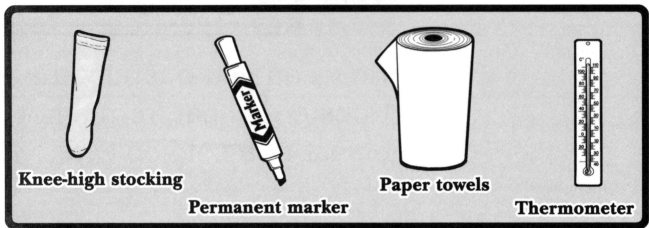

Knee-high stocking

Permanent marker

Paper towels

Thermometer

Do this:

1. Stuff paper towels in the stocking until you have a nice, even snake body.

2. Put the thermometer inside the stocking. Make sure you can read the numbers.

3. Tie a knot at the end of the stocking to hold the paper towels inside. You can make lots of knots to make it look like a rattle on the snake's tail.

4. Use the permanent marker to decorate your snake.

5. If necessary, adjust the position of the thermometer so that it is on the snake's side. You need to be able to read it when the snake is resting on its belly, but it should not be on top of the snake.

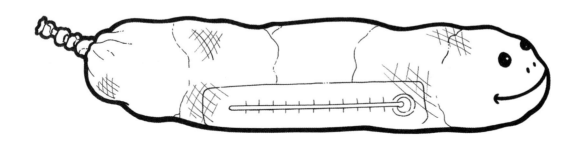

 © 2010 AIMS Education Foundation

Snake Warmers

1. Find the temperature of your snake. Record it in the table below in the space for zero minutes.

2. Fill your bottle with hot tap water all the way to the top. Screw on the lid. Make sure it doesn't leak. Dry the outside of the bottle with paper towels.

3. Place your bottle on its side. This is your rock. Put your snake on the rock.

4. After two minutes, find and record the temperature of your snake.

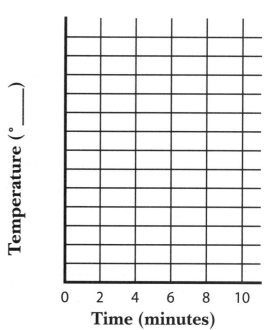

5. Repeat every two minutes for 10 minutes.

6. Make a line graph of your data. Decide on a scale for the temperature data and write in the numbers.

Time (minutes)	Temperature (°_____)
0	
2	
4	
6	
8	
10	

7. What does your graph tell you?

Temperature (°_____)

Time (minutes)

0 2 4 6 8 10

 © 2010 AIMS Education Foundation

Connecting Learning

1. What is heat conduction?

2. What are some ways that you experience heat conduction in your day-to-day life?

3. Does experiencing heat conduction always have to be you being warmed by something hotter than your body? Explain.

4. What was the temperature of your group's snake before you put it on the rock?

5. What happened to the snake's temperature after two minutes? ...four minutes? ...etc.?

6. How did your data compare to the data collected by other groups? What might be some reasons for any differences?

 © 2010 AIMS Education Foundation

Connecting Learning

7. How did your snake feel when you picked it up after it had been on the rock for 10 minutes?

8. What does this tell you about what was happening?

9. Do you think the temperature of the snake would have continued to go up if we had continued to measure for more than 10 minutes? Explain.

10. What would happen if we left our snakes on the rocks overnight?

11. What are you wondering now?

All Wrapped Up

Topic
Heat insulation

Key Question
What can we use to make a jar of water retain heat as long as possible?

Learning Goals
Students will:
- compare an insulated jar of heated water to one that is not insulated, and
- conduct a tournament to determine the best insulated jar.

Guiding Documents
Project 2061 Benchmarks
- *Recognize when comparisons might not be fair because some conditions are not kept the same.*
- *Some materials conduct heat much better than others. Poor conductors can reduce heat loss.*
- *When warmer things are put with cooler ones, the warm ones lose heat and the cool ones gain it until they are all at the same temperature. A warmer object can warm a cooler one by contact or at a distance.*
- *People try to conserve energy in order to slow down the depletion of energy resources and/or to save money.*

NRC Standards
- *Employ simple equipment and tools to gather data and extend the senses.*
- *Heat can be produced in many ways, such as burning, rubbing, or mixing one substance with another. Heat can move from one object to another by conduction.*
- *Heat moves in predictable ways, flowing from warmer objects to cooler ones, until both reach the same temperature.*

NCTM Standards 2000*
- *Select and apply appropriate standard units and tools to measure length, area, volume, weight, time, temperature, and the size of angles*
- *Collect data using observations, surveys, and experiments*
- *Represent data using tables and graphs such as line plots, bar graphs, and line graphs*

Math
Measurement
   temperature
   time
Graphing

Science
Physical science
   heat energy
      insulation

Integrated Processes
Observing
Controlling variables
Collecting and recording data
Comparing and contrasting
Interpreting data

Materials
For each group:
   2 baby food jars
   2 matching thermometers

For the class:
   various kinds of insulating materials (see
      Management 1)
   source to heat water (see *Management 2*)
   clock (kitchen timer advisable for *Part Two*)
   transparent or masking tape

Background Information
Conservation of energy is an important concern today, both because of depleting energy supplies and because of the cost to the consumer. The water heater is one of the major energy users in a home. How can it be made more energy efficient?

The way many water heaters are now constructed has made them more energy efficient. They are glass-lined; glass is a good insulator. Between the glass lining and the outer shell is a foam insulation blanket, commonly with an R-16 rating. Some high-efficiency water heaters have even greater insulation. To improve the efficiency of older water heaters without foam insulation, insulation blankets can be wrapped around the outside. Insulation slows the cooling of the water so the heating element does not have to be used as often.

It is recommended that the water heater be set at 120°F, or between 100°F and 120°F in households with young children. However, homes with dishwashers should be set at 140°F.* The higher the thermostat is set, the more energy will be needed.

There are two major conclusions to be drawn from this investigation: *insulation is better than no insulation,* and *the kind of insulation matters.* Some materials are conductors rather than insulators. Some insulators are better than others. Trapped air, such as that found in crumpled newspaper, is a good heat insulator.

These findings can be applied to water heaters, water pipes exposed to cold weather, cups used for hot beverages, walls within buildings, and people trying to keep warm.

*Pacific Gas and Electric Company

Management

1. Gather a variety of materials for students to try such as newspapers, cotton batting, wool socks, foam peanuts, foil, down mittens, straw, dirt, etc. If you wish, give students the challenge and allow them a day or two to collect their own materials. Encourage creativity in their choices and their construction of the wrap by not offering a finished example.

2. A hot plate operated by an adult can be used to heat water. Due to safety concerns, water should not be heated above 70°C (about 160°F).

3. There are two options for doing this activity: 1) Students can use the first page to plan their own investigation (see *Guided Planning*) or 2) use the graph and tournament pages to do a two-part investigation.

4. Groups of two to four are suggested.

5. For a fair test, several variables need to be controlled. The thermometers within a group must have matching readings, readings need to be taken at consistent intervals, and the starting water temperature must be the same within each group. To reduce the amount of water exposed directly to the air, consider covering the tops of both jars with something such as foil that can be slit to allow a thermometer to be inserted.

6. For the first part, the same starting temperature will allow comparisons between the insulated and non-insulated jars belonging to individual groups. Because it is difficult to keep starting temperatures the same from group to group, direct comparisons between groups cannot be made at this point. However, try to keep the starting temperatures as close as possible to each other so the insulated jars can be accurately seeded for the tournament.

7. Since water cools quickly when removed from the heat source and the goal is to have similar starting temperatures, pour water in the jars of two or three groups at a time. Then reheat it, check the temperature and pour for the next groups. Because of the staggered start, groups will be taking temperature readings at different times and will need to monitor their own time intervals.

8. Once the water is poured, the gathering of temperature data must begin immediately.

9. If there are more than eight groups for the tournament, make a second copy of the tournament brackets and tape it below the first.

(The following is for those students ready for more independent work.)

Guided Planning: A page is included for small groups to use in planning the investigation. After initial small group discussion, the class will need to agree on the time intervals to measure temperature and how they will conduct a fair test. A significant variable to be controlled is the starting temperature.

Procedure
Part One

1. Ask, "In what ways do we use heated water?" [washing and rinsing dishes, taking baths or showers, washing clothes, making hot drinks like tea or coffee, heating swimming pools, etc.] "Do we ever need to be concerned about keeping water warm? Explain." [Yes. So the hot chocolate or coffee stays warm long enough to drink. So the water in pipes doesn't freeze and cause damage. So energy is not wasted.]

2. Have students identify the source of heated water in our homes. [water heater] Explain that the water heater is one of the biggest users of energy in a home. To conserve energy and save money, we want the water to stay as warm as possible so the water heater will not have to go on so often. Tell students that they will be using baby food jars to represent water heaters and investigate some ways that might help keep water warm.

3. Give students the graph page and review the rules. Discuss what the class definition of "wrap" will be. Will only the side of the jar be covered? ...the bottom and side? ...the bottom, side, and top? For the latter choice, the top must be removable to allow the water to be poured.

4. Determine together how often temperature readings should be taken. (Every five minutes is suggested.) Have students label the horizontal axis of the graph, starting with *zero.*

5. Direct each group to use the collected materials to wrap one of the jars.

6. Pour heated water into the jars of two or three groups. Be careful not to wet the insulating material, protecting it with waxed paper or plastic wrap if necessary. Instruct students to take an *immediate* reading of the starting temperature, then watch the clock and take regular readings for the designated number of minutes. (For five-minute intervals, readings will be taken for a total of 30 minutes.)

7. Heat the water again and repeat for two or three more groups. Continue until all groups have performed the investigation.
8. Ask, "How shall we number the *Degrees Celsius* part of the graph?" [Survey the range of temperature readings—highest and lowest—in the class.] Have students number the graph in one-degree intervals, starting with the lowest temperature reading.
9. Discuss the questions for *Part One*.

Part Two
1. Set up the tournament brackets using each group's chosen name. Seed the top four insulators from *Part One*, putting one in each pair of brackets.
2. Have students line up one or two competing pairs of insulated jars in the back of the room.
3. Pour the heated water in the jars. Have students record the starting temperature on the tournament chart, set a kitchen timer for 30 minutes, and record the ending temperature after each group's name. Write the winner's name in the next round of brackets.
4. Repeat until all the pairs have competed and a final winner is determined.
5. Discuss the tournament results.

Connecting Learning
Part One
1. How is the conduction of heat evident in this lesson?
2. What is the purpose of insulation?
3. What did you learn about insulators? Use data to defend your answer.
4. How can we compare the wraps from different groups? Can we make comparisons from the data we already have?

Part Two
1. How did we make sure this was a fair test? [The starting temperature for those in direct competition was kept the same, matching thermometers were used, etc.]
2. How does the effectiveness of the wraps compare? What conclusions can you draw?
3. How could this information be related to a water heater? [The heater should be insulated; the newer ones are. An older water heater can be wrapped with an insulation blanket to help reduce energy costs.]
4. What other things need to be insulated?
5. What are you wondering now?

Extension
Research other parts of the house that are insulated. [walls, ceiling, etc.] Why would a house need to be insulated?

Home Links
1. Have students, with parental supervision, check for evidence of insulation on their water heater. If the water heater is older, an insulation blanket might be considered. New water heaters have built-in insulation features to promote energy savings. It may even say on the tank how costly the water heater is to run.
2. Have students, with parental supervision, find the number of kilowatts used by their water heater. Use the local utility rates and the estimated number of hours the heater operates to calculate how much it costs to run per month.

* Reprinted with permission from *Principles and Standards for School Mathematics,* 2000 by the National Council of Teachers of Mathematics. All rights reserved.

All Wrapped Up

Key Question

What can we use to make a jar of water retain heat as long as possible?

Learning Goals

Students will:

- compare an insulated jar of heated water to one that is not insulated, and

- conduct a tournament to determine the best insulated jar.

 © 2010 AIMS Education Foundation

All Wrapped Up

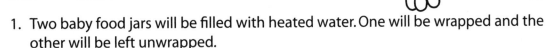

Challenge: Wrap a jar to retain heat as long as possible.

Rules

1. Two baby food jars will be filled with heated water. One will be wrapped and the other will be left unwrapped.
2. The diameter of the jar and wrapping can be no larger than 15 cm.
3. Thermometers must be able to be inserted into the jars and read without removing them from the water.

Planning:

1. Brainstorm wrapping materials you might use. Will you use one material or a combination of materials?

*2. When and how often will you read the temperature?

3. How will you record the results?

4. What is the most appropriate way to show the results?

*5. How will a fair test be made so that the wrapped-jar results of all the groups can be compared?

*After discussion within groups, all groups need to come to agreement on a common procedure.

All Wrapped Up

Challenge: Wrap a jar to retain heat as long as possible.

Rules

1. Two baby food jars will be filled with heated water. One will be wrapped and the other will be left unwrapped.
2. The diameter of the jar and wrapping can be no larger than 15 cm.
3. Thermometers must be able to be inserted into the jars and read without removing them from the water.

Record and graph the data from both jars.

Which had the smallest change in temperature?

Key
- ☐ wrapped
- ☐ unwrapped

Temperature (°C)

Elapsed Time (Minutes)

All Wrapped Up

Seed the top four insulators for the tournament, putting one in each pair of brackets. Record the beginning temperature on the line and the ending temperature after each groups' name.

_____ Beginning temperature

_____ Beginning temperature

_____ Beginning temperature

_____ Beginning temperature

_____ Beginning temperature

_____ Beginning temperature

_____ Beginning temperature

All Wrapped Up

Connecting Learning

Part One

1. How is the conduction of heat evident in this lesson?

2. What is the purpose of insulation?

3. What did you learn about insulators? Use data to defend your answer.

4. How can we compare the wraps from different groups? Can we make comparisons from the data we already have?

All Wrapped Up

Connecting Learning

Part Two

1. How did we make sure this was a fair test?

2. How does the effectiveness of the wraps compare? What conclusions can you draw?

3. How could this information be related to a water heater?

4. What other things need to be insulated?

5. What are you wondering now?

EVENING Out Temperatures

Topic
Temperature

Key Question
How will the temperatures of three different samples of water change over time?

Learning Goals
Students will:
- find the temperatures of three different samples of water—hot water, room temperature water, and cold water;
- measure the temperature of the samples every 15 minutes;
- determine how long it takes for each sample to reach room temperature; and
- graph and analyze their data.

Guiding Documents
Project 2061 Benchmarks
- *When warmer things are put with cooler ones, the warm ones lose heat and the cool ones gain it until they are all at the same temperature. A warmer object can warm a cooler one by contact or at a distance.*
- *Tables and graphs can show how values of one quantity are related to values of another.*

NRC Standards
- *Heat moves in predictable ways, flowing from warmer objects to cooler ones, until both reach the same temperature.*
- *Objects have many observable properties, including size, weight, shape, color, temperature, and the ability to react with other substances. Those properties can be measured using tools, such as rulers, balances, and thermometers.*
- *Employ simple equipment and tools to gather data and extend the senses.*

NCTM Standards 2000*
- *Select and apply appropriate standard units and tools to measure length, area, volume, weight, time, temperature, and the size of angles*
- *Collect data using observations, surveys, and experiments*
- *Represent data using tables and graphs such as line plots, bar graphs, and line graphs*

Math
Measurement
  temperature
  elapsed time
Graphing
  bar graphs

Science
Physical science
  heat energy

Integrated Processes
Observing
Predicting
Collecting and recording data
Comparing and contrasting
Analyzing data

Materials
Immersion thermometers (see *Management 5*)
Styrofoam cups
Water pitchers
Hot water (see *Management 2*)
Cold water (see *Management 3*)
Crayons
Student pages

Background Information
Temperature is a measure of the amount of heat something has; therefore, a change in temperature indicates a change in heat. Higher temperatures indicate more heat than lower temperatures.

When a cup of hot liquid is left to sit, it will eventually reach room temperature. This is due to loss of heat energy to the surrounding air. Conversely, a cup of cold liquid will reach room temperature by gaining heat energy from the surrounding air. Both of these changes can be measured quantitatively using thermometers. Depending on the kind of cup, amount of water used, and initial temperature of the water, it can take an hour or more before equilibrium is reached.

Management
1. Each group of students needs three immersion thermometers. Students can work in groups of three to five.
2. Use the hottest tap water possible. Be sure to get it right before doing the activity or keep it in an insulated container so that it is still hot when students get it. Caution students to handle it carefully.

3. The day before doing the activity, fill a large pitcher with water and put it in the refrigerator so that it is completely chilled. Do not bring it out until immediately prior to pouring the water in students' cups. Fill a second large pitcher with tap water and allow it to sit for several hours so that it is at room temperature.

4. When doing the activity, be sure that all groups' cups are in the same kind of conditions. For example, be sure none are in direct sunlight or in the direct path of a fan or anything else that would change how quickly they reach room temperature.

5. If you are using thermometers with dual scales, you may want to put masking tape over the scale you want students to ignore. Thermometers are available from AIMS (item number 1976).

6. Line graphs are normally used to show change over time. This activity uses bar graphs for their simplicity. If desired, the graph can be altered for the line graph format.

Procedure

1. Ask students if they have ever left a cool drink outside on a hot summer day. Invite them to share what happened to the drink. [After a while, any ice melts and the drink warms up.]

2. Have students think about drinking hot cocoa or cider in the winter. Ask them if there is anything they usually do before drinking hot drinks. [Perhaps they blow on the drink or let it cool off for a few minutes.]

3. Challenge students to explain what these two scenarios have in common. Discuss how in both cases, the drinks eventually come to the same temperature as the air. In one case, a cool drink is warming up, and in the other, a hot drink is cooling down.

4. Ask students to make predictions, based on their prior experiences, as to how long it will take a hot drink to cool down and a cool drink to warm up. Distribute the first two student pages and have students record their predictions in the spaces provided.

5. Explain that students are going to have the opportunity to see how the temperatures of three samples of water change over time.

6. Have students get into their groups and give each group three Styrofoam cups and three thermometers. Instruct students to label each cup using a crayon. One cup is for cold water, one cup is for hot water, and one cup is for room temperature water.

7. Review the proper way to read the temperature on a thermometer, and have students practice by reading the current room temperature. Tell students whether you would like them to make their measurements using Fahrenheit or Celsius.

8. Invite two students from each group to come to the place where you have the water pitchers with the room temperature and cold water. Fill each group's cups with the two kinds of water (be sure to check the labels to make sure you're filling the correct cup). The cups do not need to be full—the more water in the cup, the longer it will take to reach room temperature—but all should have the same amount of water.

9. When all students have returned to their groups, move from group to group and fill the remaining cups with the hot water.

10. Have students measure and record the temperature of the water in each cup. (They should allow enough time (approximately three minutes) for the thermometer to adjust to the temperature before reading each.) Have them record the current time in the table above the first measurement and draw hands on the clocks to show the starting time.

11. Every 15 minutes, have the groups measure and record the temperatures and the time. Continue this until all of the cups have reached approximately the same temperature. (If this takes more than an hour and 45 minutes, students will have to add columns to the data table and the graphs.)

12. Once measurements are completed, have students draw hands on the clocks to show the time at which each sample reached room temperature (these times will likely be different). Have them calculate and record the actual amount of time it took for the water to reach room temperature and answer the questions at the bottom of the second page.

13. Distribute the third student page and assist students in labeling their graphs and deciding on the appropriate scale for the temperature. Have them make bar graphs using their groups' data.

14. Discuss the differences between the hot and cold cups as well as the differences among the groups.

Connecting Learning

1. How hot was your hot water when you started? How cold was your cold water? Was there a greater temperature difference between the hot water and the room temperature water or the cold water and the room temperature water?

2. How long did you predict it would take the cold water to warm up? How long did it actually take? Was this longer or shorter than what you predicted?

3. How long did you predict it would take the hot water to cool down? How long did it actually take? Was this longer or shorter than what you predicted?

4. Did it take longer for the cool water to warm up or for the hot water to cool down (or did both happen in the same amount of time)? Did this surprise you? Why or why not?

5. Were the temperature measurements the same from group to group at any given time? What might be some reasons for any differences?

6. What do the graphs show you? Which cup changed temperature most quickly? How do you know?

7. What do you think would have happened to the time it took for the cold water to warm up if we had added ice? Why?

8. What do you think would have been different about our results if we had done this experiment outside in the sun on a warm summer day? What if we had done it outside on a cold winter day?

9. Why do you think the temperatures of the hot and cold water changed? [The cold water was gaining heat from the surrounding air. The hot water was losing heat to the surrounding air.]

10. What are some examples of warm water cooling and cold water warming? [Swimming pools, oceans, lakes, rivers, ponds, etc., all change temperature with the seasons.]

11. What are you wondering now?

Extensions
1. Try adding ice to the cold water cup to see how much longer it takes to reach room temperature.

2. Try using different kinds of cups (paper, ceramic mugs, etc.) to see how that affects the amount of time it takes the liquid to reach room temperature.

3. Increase (or decrease) the amount of liquid and see how much slower (or faster) the water reaches room temperature.

4. Do the activity outdoors to see how the air temperature changes the results.

* Reprinted with permission from *Principles and Standards for School Mathematics*, 2000 by the National Council of Teachers of Mathematics. All rights reserved.

 © 2010 AIMS Education Foundation

EVENING Out Temperatures

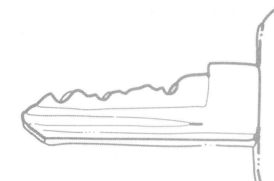

Key Question

How will the temperatures of three different samples of water change over time?

Learning Goals

Students will:

- find the temperatures of three different samples of water—hot water, room temperature water, and cold water;
- measure the temperature of the samples every 15 minutes;
- determine how long it takes for each sample to reach room temperature; and
- graph and analyze their data.

EVENING Out Temperatures

1. Make your predictions.
2. Draw hands on the first clock to show the start time.
3. When the water reaches room temperature, draw hands on the second clock to show the end time.
4. Find and record the elapsed time.

How long will it take cold water to reach room temperature?

I predict: _____

Actual time: _____

Start Time

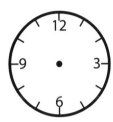

End Time

How long will it take hot water to reach room temperature?

I predict: _____

Actual time: _____

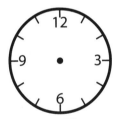

Start Time

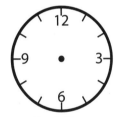

End Time

© 2010 AIMS Education Foundation

Evening Out Temperatures

Record the water temperature in your cups. Write the time in the table. Measure and record every 15 minutes. Continue until all three cups are the same temperature.

Time								
Temperature — Cold water								
Room temperature water								
Hot water								

1. How did your predictions compare to the actual results? Why do you think this is?

2. Did it take longer for the hot water to cool down or for the cold water to warm up? How much longer?

3. Why do you think the water changed temperature?

© 2010 AIMS Education Foundation

EVENING Out Temperatures

Graph your results. Label each graph with the correct temperature (°C or °F). Write in the appropriate numbers. Use the same numbers on each graph.

Cold Water

Temperature (°____)

Elapsed Time (Minutes)
0 15 30 45 60 75 90 105

Room Temperature Water

Temperature (°____)

Elapsed Time (Minutes)
0 15 30 45 60 75 90 105

Hot Water

Temperature (°____)

Elapsed Time (Minutes)
0 15 30 45 60 75 90 105

What do the graphs show you?

EVENING OUT TEMPERATURES

Connecting Learning

1. How hot was your hot water when you started? How cold was your cold water? Was there a greater temperature difference between the hot water and the room temperature water or the cold water and the room temperature water?

2. How long did you predict it would take the cold water to warm up? How long did it actually take? Was this longer or shorter than what you predicted?

3. How long did you predict it would take the hot water to cool down? How long did it actually take? Was this longer or shorter than what you predicted?

          © 2010 AIMS Education Foundation

EVENING Out Temperatures

Connecting Learning

4. Did it take longer for the cool water to warm up or for the hot water to cool down? Did this surprise you? Why or why not?

5. Were the temperature measurements the same from group to group at any given time? What might be some reasons for any differences?

6. What do the graphs show you? Which cup changed temperature most quickly? How do you know?

7. What do you think would have happened to the time it took for the cold water to warm up if we had added ice? Why?

 © 2010 AIMS Education Foundation

EVENING Out Temperatures

Connecting Learning

8. What do you think would have been different about our results if we had done this experiment outside in the sun on a warm summer day? What if we had done it outside on a cold winter day?

9. Why do you think the temperatures of the hot and cold water changed?

10. What are some examples of warm water cooling and cold water warming?

11. What are you wondering now?

 © 2010 AIMS Education Foundation

When Hot and Cold Meet

Topic
Convection currents

Key Question
What happens when a small container of colored hot water and a small container of colored cold water are placed in a large container of room temperature water?

Learning Goals
Students will:
- observe the action of hot water currents placed in room temperature water,
- observe the action of cold water currents placed in room temperature water, and
- apply the movements of the currents to convection currents that cause air to move.

Guiding Documents
Project 2061 Benchmark
- *When warmer things are put with cooler ones, the warm ones lose heat and the cool ones gain it until they are all at the same temperature. A warmer object can warm a cooler one by contact or at a distance.*

NRC Standard
- *Heat moves in predictable ways, flowing from warmer objects to cooler ones, until both reach the same temperature.*

Science
Physical science
    heat energy
        convection

Integrated Processes
Observing
Communicating
Predicting
Applying

Material
Large transparent tub or small aquarium
2 baby food jars
Red and blue food coloring
Aluminum foil
2 rubber bands
Hot water
Cold water
Room temperature water
Student page

Background Information
Heat energy tends to move from warmer objects to cooler objects. In convection, a portion of a fluid becomes hotter (or colder) than the rest of the fluid. Its density becomes different from the rest of the fluid. Hotter fluid is generally less dense than the rest, so it rises; colder fluid is generally more dense, so it sinks. Eventually, the fluid will all come to the same temperature.

In this activity, the hot colored water in the small jar is less dense than the surrounding cooler water. The hot water is buoyed up or pushed upward by the cooler, denser water. It rises and resembles smoke pouring from a chimney. The hot colored water tends to rise and then rest along the top or surface of the cooler water. If it is left long enough and the water is not disturbed, the colored water will then cool and mix with the other water.

The cold water is more dense than the surrounding water, so the cold water settles to the bottom of the container. Some students may have experienced a temperature difference in a swimming pool or lake. As they dive toward the bottom, the water is cooler than the surface water that is less dense.

Air is a fluid and will behave like the currents of water. The uneven heating of the Earth causes the movement of air that results in changes in weather patterns across the Earth.

Management
1. Use caution when handling the hot water. It may be advisable to use this as a teacher-led demonstration.
2. The large container should be filled with room temperature water.
3. The greater the difference in the temperature between the hot and cold water, the more dramatic the results will be.

Procedure
1. Fill an aquarium or large transparent container with room temperature water.
2. Fill one small jar with hot water and add two or three drops of red food coloring. Cover the jar opening with aluminum foil and put a rubber band around the neck.
3. Gently lower it into the large container or aquarium.
4. Puncture the aluminum foil with two small holes opposite each other. Place the small jar on its side as illustrated.
5. Have students observe what happens and make a record on their student pages.

6. Fill the other small jar with ice water and add two or three drops of blue food coloring. Cover the jar with aluminum foil and put a rubber band around the neck.

7. Gently lower it into the large container or aquarium. Puncture the aluminum foil with two small holes opposite each other. Place the small jar on its side as illustrated.

8. Have the students observe what is happening and draw the results on the student page.

9. Have the students write their observations on the bottom of the student page.

Connecting Learning

1. What did you notice happened to the hot water from the baby food jar? Why do you think this happened? [The hot water is less dense and so rises.]

2. What happened to the cold water? Why? [The cold water is more dense so it sank to the bottom.]

3. What would happen if the colored water in the container were left alone for one hour? [If the water is not disturbed, the hot and cold water will mix as their temperatures equalize and the colors will blend.]

4. What other material rises as it is heated? [air]

5. What happens when air moves? [Wind is created.]

6. Explain in your own words what is meant when we say that uneven heating of the Earth causes air movements.

7. What are you wondering now?

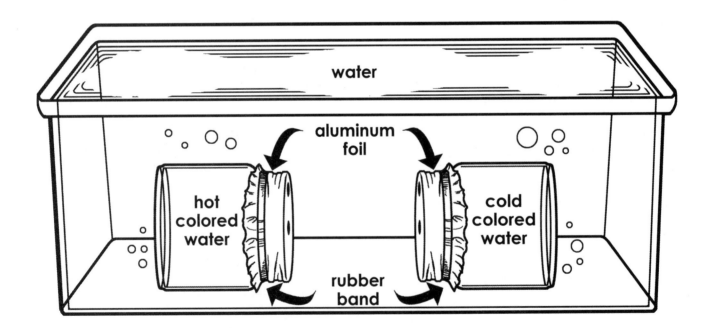

When Hot and Cold Meet

Key Question

What happens when a small container of colored hot water and a small container of colored cold water are placed in a large container of room temperature water?

Learning Goals

Students will:

- observe the action of hot water currents placed in room temperature water,

- observe the action of cold water currents placed in room temperature water, and

- apply the movements of the currents to convection currents that cause air to move.

 © 2010 AIMS Education Foundation

When Hot and Cold Meet

1. Put the jar of hot colored water into the tank of water. Use a pencil to make two holes in the aluminum foil. Turn the jar on its side.

2. Put the jar of cold colored water into the tank of water. Use a pencil to make two holes in the aluminum foil. Turn the jar on its side.

water

aluminum foil

hot colored water

cold colored water

rubber band

3. Write about what you saw happen.

4. Why do you think this happened?

     © 2010 AIMS Education Foundation

When Hot and Cold Meet

Connecting Learning

1. What did you notice happened to the hot water from the baby food jar? Why do you think this happened?

2. What happened to the cold water? Why?

3. What would happen if the colored water in the container were left alone for one hour?

4. What other material rises as it is heated?

5. What happens when air moves?

6. Explain in your own words what is meant when we say that uneven heating of the Earth causes air movements.

7. What are you wondering now?

          © 2010 AIMS Education Foundation

Energy Explorations: Sound, Light, and Heat

Materials List

Equipment

*Tuning forks (#1973 or #1974)
*Table tennis ball (#1975)
*Metal Slinkys® (#1981)
*Meter stick (#1908)
*Metric rulers (#1909)
*Battery holders (#1960)
*Bulb holders (#1958)
*Bulbs (#1962)
*Insulated wire (#1968)
*Mirrors (#1979)

*Mini hand lenses (#3072)
*Hinged mirrors (#1987)
*Kaleidoscope kits (#4135)
*Liter boxes (#1913)
*Prisms (#1978)
*Thermometers (#1976)
Wind-up clock
Metal fork or serving spoons
Paper cutter
Wooden rulers

Calculators
Flashlights
Overhead projector
Safety glasses
Hair dryer
Desk lamp
Large transparent tub or small
    aquarium

Consumables and Non-consumables

Rubber bands, various sizes
Transparent tape
Masking tape
White glue
Pencils
Crayons
Colored pencils
Permanent markers
Paper clips
Pushpins
Metal scissors
Hole punches
Card stock
Construction paper—red,
    yellow, blue, black, white
White paper, 12" x 18"
Index cards, 4" x 6"
Transparency film
Old newspaper
Tissue paper
Cardboard

Styrofoam cups
Plastic cups—3 oz, 9 oz, 10 oz
Paper cups
Paper bag
Large zipper-type plastic bag
Paper plates
Paper towels
Clear plastic bottles
1-gallon water bottles
Aluminum foil
Wax paper
Empty metal coffee can
Glass jar
Baby food jars
Five identical glass bottles
Shoebox
Toilet paper tubes
Wine corks
Half-pint milk cartons
Drinking straw
Water pitchers

Fluorescent bulb covers
D cells
Clay
Craft sticks
Googly eyes
Cloth
Crochet thread
Sewing thread
String
Metal coat hangers
Shiny pennies
Candle
Matches
Blocks of wood
Sandpaper
Hand lotion
Knee-high stockings
Food coloring—red, blue,
    yellow, green, orange
Cooking oil
Water, hot and cold

*Available from AIMS

The AIMS Program

AIMS is the acronym for "Activities Integrating Mathematics and Science." Such integration enriches learning and makes it meaningful and holistic. AIMS began as a project of Fresno Pacific University to integrate the study of mathematics and science in grades K-9, but has since expanded to include language arts, social studies, and other disciplines.

AIMS is a continuing program of the non-profit AIMS Education Foundation. It had its inception in a National Science Foundation funded program whose purpose was to explore the effectiveness of integrating mathematics and science. The project directors, in cooperation with 80 elementary classroom teachers, devoted two years to a thorough field-testing of the results and implications of integration.

The approach met with such positive results that the decision was made to launch a program to create instructional materials incorporating this concept. Despite the fact that thoughtful educators have long recommended an integrative approach, very little appropriate material was available in 1981 when the project began. A series of writing projects ensued, and today the AIMS Education Foundation is committed to continuing the creation of new integrated activities on a permanent basis.

The AIMS program is funded through the sale of books, products, and professional-development workshops, and through proceeds from the Foundation's endowment. All net income from programs and products flows into a trust fund administered by the AIMS Education Foundation. Use of these funds is restricted to support of research, development, and publication of new materials. Writers donate all their rights to the Foundation to support its ongoing program. No royalties are paid to the writers.

The rationale for integration lies in the fact that science, mathematics, language arts, social studies, etc., are integrally interwoven in the real world, from which it follows that they should be similarly treated in the classroom where students are being prepared to live in that world. Teachers who use the AIMS program give enthusiastic endorsement to the effectiveness of this approach.

Science encompasses the art of questioning, investigating, hypothesizing, discovering, and communicating. Mathematics is a language that provides clarity, objectivity, and understanding. The language arts provide us with powerful tools of communication. Many of the major contemporary societal issues stem from advancements in science and must be studied in the context of the social sciences. Therefore, it is timely that all of us take seriously a more holistic method of educating our students. This goal motivates all who are associated with the AIMS Program. We invite you to join us in this effort.

Meaningful integration of knowledge is a major recommendation coming from the nation's professional science and mathematics associations. The American Association for the Advancement of Science in *Science for All Americans* strongly recommends the integration of mathematics, science, and technology. The National Council of Teachers of Mathematics places strong emphasis on applications of mathematics found in science investigations. AIMS is fully aligned with these recommendations.

Extensive field testing of AIMS investigations confirms these beneficial results:

1. Mathematics becomes more meaningful, hence more useful, when it is applied to situations that interest students.
2. The extent to which science is studied and understood is increased when mathematics and science are integrated.
3. There is improved quality of learning and retention, supporting the thesis that learning which is meaningful and relevant is more effective.
4. Motivation and involvement are increased dramatically as students investigate real-world situations and participate actively in the process.

We invite you to become part of this classroom teacher movement by using an integrated approach to learning and sharing any suggestions you may have. The AIMS Program welcomes you!

© 2010 AIMS Education Foundation

AIMS Education Foundation Programs

When you host an AIMS workshop for elementary and middle school educators, you will know your teachers are receiving effective, usable training they can apply in their classrooms immediately.

AIMS Workshops are Designed for Teachers
- Correlated to your state standards;
- Address key topic areas, including math content, science content, and process skills;
- Provide practice of activity-based teaching;
- Address classroom management issues and higher-order thinking skills;
- Give you AIMS resources; and
- Offer optional college (graduate-level) credits for many courses.

AIMS Workshops Fit District/Administrative Needs
- Flexible scheduling and grade-span options;
- Customized (one-, two-, or three-day) workshops meet specific schedule, topic, state standards, and grade-span needs;
- Prepackaged four-day workshops for in-depth math and science training available (includes all materials and expenses);
- Sustained staff development is available for which workshops can be scheduled throughout the school year;
- Eligible for funding under the Title I and Title II sections of No Child Left Behind; and
- Affordable professional development—consecutive-day workshops offer considerable savings.

University Credit—Correspondence Courses
AIMS offers correspondence courses through a partnership with Fresno Pacific University.
- Convenient distance-learning courses—you study at your own pace and schedule. No computer or Internet access required!

Introducing AIMS State-Specific Science Curriculum
Developed to meet 100% of your state's standards, AIMS' State-Specific Science Curriculum gives students the opportunity to build content knowledge, thinking skills, and fundamental science processes.
- Each grade-specific module has been developed to extend the AIMS approach to full-year science programs. Modules can be used as a complete curriculum or as a supplement to existing materials.
- Each standards-based module includes math, reading, hands-on investigations, and assessments.

Like all AIMS resources, these modules are able to serve students at all stages of readiness, making these a great value across the grades served in your school.

For current information regarding the programs described above, please complete the following form and mail it to: P.O. Box 8120, Fresno, CA 93747.

Information Request

Please send current information on the items checked:

____ *Basic Information Packet* on AIMS materials

____ Hosting information for AIMS workshops

____ AIMS State-Specific Science Curriculum

Name: _____

Phone:_____ E-mail:_____

Address: _____
         Street   City   State   Zip

© 2010 AIMS Education Foundation

AIMS™ Magazine

Your K-9 Math and Science Classroom Activities Resource

The AIMS Magazine is your source for standards-based, hands-on math and science investigations. Each issue is filled with teacher-friendly, ready-to-use activities that engage students in meaningful learning.

- *Four issues each year (fall, winter, spring, and summer).*

Current issue is shipped with all past issues within that volume.

| 1824 | Volume | XXV | 2010-2011 | $19.95 |
| 1825 | Volume | XXVI | 2011-2012 | $19.95 |

Two-Volume Combinations

| M21012 | Volumes | XXV & XXVI | 2010-12 | $34.95 |
| M21113 | Volumes | XXVI & XXVII | 2011-13 | $34.95 |

Complete volumes available for purchase:

| 1823 | Volume | XXIII | 2008-2009 | $19.95 |
| 1824 | Volume | XXIV | 2009-2010 | $19.95 |

AIMS Online—www.aimsedu.org

To see all that AIMS has to offer, check us out on the Internet at www.aimsedu.org. At our website you can preview and purchase AIMS books and individual activities, learn about State-Specific Science and Essential Math, explore professional development workshops and online learning opportunities, search our activities database, buy manipulatives and other classroom resources, and download free resources including articles, puzzles, and sample AIMS activities.

AIMS E-mail Specials
While visiting the AIMS website, sign up for our FREE e-mail newsletter with monthly subscriber-only specials. You'll also receive advance notice of new products.

Sign up today!

Subscribe to the AIMS Magazine

$19.95 a year!

AIMS Magazine is published four times a year.

Subscriptions ordered at any time will receive all issues for that year.

Call **1.888.733.2467** or go to **www.aimsedu.org**

© 2010 AIMS Education Foundation

AIMS Program Publications

Actions With Fractions, 4-9
The Amazing Circle, 4-9
Awesome Addition and Super Subtraction, 2-3
Bats Incredible! 2-4
Brick Layers II, 4-9
The Budding Botanist, 3-6
Chemistry Matters, 5-7
Counting on Coins, K-2
Cycles of Knowing and Growing, 1-3
Crazy About Cotton, 3-7
Critters, 2-5
Earth Book, 6-9
Electrical Connections, 4-9
Energy Explorations: Sound, Light, and Heat, 3-5
Exploring Environments, K-6
Fabulous Fractions, 3-6
Fall Into Math and Science*, K-1
Field Detectives, 3-6
Finding Your Bearings, 4-9
Floaters and Sinkers, 5-9
From Head to Toe, 5-9
Glide Into Winter With Math and Science*, K-1
Gravity Rules! 5-12
Hardhatting in a Geo-World, 3-5
Historical Connections in Mathematics, Vol. I, 5-9
Historical Connections in Mathematics, Vol. II, 5-9
Historical Connections in Mathematics, Vol. III, 5-9
It's About Time, K-2
It Must Be A Bird, Pre-K-2
Jaw Breakers and Heart Thumpers, 3-5
Looking at Geometry, 6-9
Looking at Lines, 6-9
Machine Shop, 5-9
Magnificent Microworld Adventures, 6-9
Marvelous Multiplication and Dazzling Division, 4-5
Math + Science, A Solution, 5-9
Mathematicians are People, Too
Mathematicians are People, Too, Vol. II
Mostly Magnets, 3-6
Movie Math Mania, 6-9
Multiplication the Algebra Way, 6-8
Out of This World, 4-8
Paper Square Geometry:
    The Mathematics of Origami, 5-12
Puzzle Play, 4-8
Popping With Power, 3-5

Positive vs. Negative, 6-9
Primarily Bears*, K-6
Primarily Earth, K-3
Primarily Magnets, K-2
Primarily Physics: Investigations in Sound, Light, and Heat Energy, K-2
Primarily Plants, K-3
Primarily Weather, K-3
Problem Solving: Just for the Fun of It! 4-9
Problem Solving: Just for the Fun of It! Book Two, 4-9
Proportional Reasoning, 6-9
Ray's Reflections, 4-8
Sensational Springtime, K-2
Sense-able Science, K-1
Shapes, Solids, and More: Concepts in Geometry, 2-3
The Sky's the Limit, 5-9
Soap Films and Bubbles, 4-9
Solve It! K-1: Problem-Solving Strategies, K-1
Solve It! 2nd: Problem-Solving Strategies, 2
Solve It! 3rd: Problem-Solving Strategies, 3
Solve It! 4th: Problem-Solving Strategies, 4
Solve It! 5th: Problem-Solving Strategies, 5
Solving Equations: A Conceptual Approach, 6-9
Spatial Visualization, 4-9
Spills and Ripples, 5-12
Spring Into Math and Science*, K-1
Statistics and Probability, 6-9
Through the Eyes of the Explorers, 5-9
Under Construction, K-2
Water, Precious Water, 4-6
Weather Sense: Temperature, Air Pressure, and Wind, 4-5
Weather Sense: Moisture, 4-5
What's Next, Volume 1, 4-12
What's Next, Volume 2, 4-12
What's Next, Volume 3, 4-12
Winter Wonders, K-2

Essential Math
Area Formulas for Parallelograms, Triangles, and Trapezoids, 6-8
Circumference and Area of Circles, 5-7
Effects of Changing Lengths, 6-8
Measurement of Prisms, Pyramids, Cylinders, and Cones, 6-8
Measurement of Rectangular Solids, 5-7
Perimeter and Area of Rectangles, 4-6
The Pythagorean Relationship, 6-8

Spanish Edition
Constructores II: Ingeniería Creativa Con Construcciones LEGO®, 4-9
The entire book is written in Spanish. English pages not included.

*   Spanish supplements are available for these books. They are only available as downloads from the AIMS website. The supplements contain only the student pages in Spanish; you will need the English version of the book for the teacher's text.

For further information, contact:
AIMS Education Foundation • P.O. Box 8120 • Fresno, California 93747-8120
www.aimsedu.org • 559.255.6396 (fax) • 888.733.2467 (toll free)

© 2010 AIMS Education Foundation

Duplication Rights

No part of any AIMS books, magazines, activities, or content—digital or otherwise—may be reproduced or transmitted in any form or by any means—including photocopying, taping, or information storage/retrieval systems—except as noted below.

Standard Duplication Rights

- A person or school purchasing AIMS activities (in books, magazines, or in digital form) is hereby granted permission to make up to 200 copies of any portion of those activities, provided these copies will be used for educational purposes and only at one school site.
- Workshop or conference presenters may make one copy of any portion of a purchased activity for each participant, with a limit of five activities per workshop or conference session.
- All copies must bear the AIMS Education Foundation copyright information.

Standard duplication rights apply to activities received at workshops, free sample activities provided by AIMS, and activities received by conference participants.

Unlimited Duplication Rights

Unlimited duplication rights may be purchased in cases where AIMS users wish to:
- make more than 200 copies of a book/magazine/activity,
- use a book/magazine/activity at more than one school site, or
- make an activity available on the Internet (see below).

These rights permit unlimited duplication of purchased books, magazines, and/or activities (including revisions) for use at a given school site.

Activities received at workshops are eligible for upgrade from standard to unlimited duplication rights.

Free sample activities and activities received as a conference participant are not eligible for upgrade from standard to unlimited duplication rights.

State-Specific Science modules are licensed to one classroom/one teacher and are therefore not eligible for upgrade from standard to unlimited duplication rights.

Upgrade Fees

The fees for upgrading from standard to unlimited duplication rights are:
- $5 per activity per site,
- $25 per book per site, and
- $10 per magazine issue per site.

The cost of upgrading is shown in the following examples:
- activity: 5 activities x 5 sites x $5 = $125
- book: 10 books x 5 sites x $25 = $1250
- magazine issue: 1 issue x 5 sites x $10 = $50

Purchasing Unlimited Duplication Rights

To purchase unlimited duplication rights, please provide us the following:
1. The name of the individual responsible for coordinating the purchase of duplication rights.
2. The title of each book, activity, and magazine issue to be covered.
3. The number of school sites and name of each site for which rights are being purchased.
4. Payment (check, purchase order, credit card)

Requested duplication rights are automatically authorized with payment. The individual responsible for coordinating the purchase of duplication rights will be sent a certificate verifying the purchase.

Internet Use

AIMS materials may be made available on the Internet if all of the following stipulations are met:
1. The materials to be put online are purchased as PDF files from AIMS (i.e., no scanned copies).
2. Unlimited duplication rights are purchased for all materials to be put online for each school at which they will be used. (See above.)
3. The materials are made available via a secure, password-protected system that can only be accessed by employees at schools for which duplication rights have been purchased.

AIMS materials may not be made available on any publicly accessible Internet site.

© 2010 AIMS Education Foundation